Handbook of
Irrigation and Drainage Systems
Management, Operation and Maintenance

Handbook of
Irrigation and Drainage Systems
Management, Operation and Maintenance

Kenneth Johnson

Editor

KOROS PRESS LIMITED
London, UK

Handbook of Irrigation and Drainage Systems: Management, Operation and
Maintenance

© 2012
Printed in 2017 for Sale in the Indian Subcontinent

Published by
Koros Press Limited
3 The Pines, Rubery B45 9FF, Rednal,
Birmingham, United Kingdom

Tel.: +44-7826-930152
Email: info@korospress.com
www.korospress.com

ISBN: 978-1-78163-008-2

Editor: Kenneth Johnson

10 9 8 7 6 5 4 3 2 1

British Library Cataloguing in Publication Data
A CIP record for this book is available from the British Library

Exclusively distributed by CBS Publishers & Distributors Pvt. Ltd.
Sales & Distribution Rights only for India, Pakistan, Bangladesh, Sri
Lanka, Nepal and Bhutan.This book is not to be sold outside these
territories.

Contents

Preface   *vii*

1. Irrigation Methods and Techniques   **1**

Methods · Irrigation Scheduling · Choosing the most Advantageous Field for Drip Irrigation · Using Drip Farming in Winter · Maintenance of Drip Irrigation Systems · Drip Irrigation Hoses Available in the Market · Advantages of a Subsurface Drip Irrigation System · Drip Irrigation Pressure Regulator · Traditional Irrigation Fails Themoney Test · Modern Irrigation Techniques · Can Irrigation be Sustainable · Three Regional Irrigation Perspectives in Rice-Based Systems · Climatic, Water Balance and Hydrological Comparison of Three Subsystems · Types of Irrigation · Surface Irrigation · Centre Pivot Irrigation · Reservoir · Types · Modelling Reservoir Management · Units of Measurement · Spate Irrigation · Where does one Find Spate Irrigation Systems? · System Captives: Change and Stagnation in Farmer-managed Water Delivery Schedules · Changing Water Delivery Schedules: Three Cases from Baluchistan · Charging for Irrigation Water by Volume-electricity would Conserve Water Resources in Greece · Constraints in Practising Irrigation Scheduling

2. Irrigation Management   **90**

Water Management · Environmental Impact of Irrigation · Reduced Downstream River Water Quality · Factors Influencing Irrigation · Partial Rootzone Drying · Tidal Irrigation · Irrigation of Alluvial Fans · Case Studies · Okavango · Drainage · Drainage in the 19th Century · Social and Ecological Effects of Tile Drainage · Drainage System (Agriculture) · Drainage System Design · Participatory Irrigation Management · Inadequate Water Availability at the Lowest Outlets · Comparative Advantage · Continuum of Involvement in Management Decisions · Training Strategy · Bylaws of the Water User Association · Organizing Processes · Restructuring the Irrigation Agency · Assessing Users' Willingness and Capacity to Pay · Linking Fees and Services · Style and Methods

• Participatory Irrigation Management: Benefits and Second Generation Problems • Process of Introducing New Forms of PIM • National Policies • The Mode of Implementing PIM • Impacts on Operational Procedures • Irrigation Agencies • Summary and Conclusions

3. **Population and Irrigation Water Management**   **200**

Population Policy: Past Record and Future Outlook • Water Policies: Managing a Challenge • Case Study of Irrigation Management • Development of Small-Scale Hydraulics in Morocco • Creation of the AUEA and their Union • Analysis, Suggestions and Questions • Agroeconomic Impacts • Operation and Maintenance • Role of Water user Groups • Irrigation Infrastructure • The Bank and the Big Bang • The Latin American Experience • Socioeconomic Issues in Irrigation Literature: Approaches, Concepts, and Meanings • Equity Among Users • Paddy Irrigation and Water Management in Southeast Asia • Agroeconomic Impacts • Operation and Maintenance • History of Irrigation Development in India • Soil, Water and Plant Characteristics Important to Irrigation • Soil Permeability and Infiltration • In General, these Soils are not Recommended for Irrigation • Irrigation Water Quality • The Interaction Between Soil and Water • How Plants get Water from Soil • What is Soil Moisture? • Why is Measuring Soil Moisture Important? • Defining the Range of Uncertainty Associated with Remotely Sensed Soil Moisture • Salinity Due to Irrigation • Material and Methods • Leaching Model (Soil) • Plants Thrive on Varying Levels of Soil Moisture

4. **Irrigation Impact on Agricultural Growth and Poverty Alleviation**   **289**

Results and Discussions • India's Groundwater Challenge • Irrigation in India • Irrigation Water Requirements • Irrigation Cropping Pattern Zones. List of Cropping Pattern Zones • Environmental Considerations in Irrigation Development • Potential Environmental Impacts of Darns and Reservoirs • The Hadejia-Nguru Wetlands • Regional Aspects of Environmental Impacts and 'Hot Spots' • Humid West Africa

Bibliography   331

Index   335

Preface

The assessment of the irrigation potential, based on soil and water resources, can only be done by simultaneously assessing the irrigation water requirements. Net irrigation water requirement is the quantity of water necessary for crop growth. It is expressed in millimetres per year or in m^3/ha per year. It depends on the cropping pattern and the climate. Information on irrigation efficiency is necessary to be able to transform NIWR into gross irrigation water requirement, which is the quantity of water to be applied in reality, taking into account water losses. Multiplying GIWR by the area that is suitable for irrigation gives the total water requirement for that area. In this study water requirements are expressed in km^3/year. Calculations of irrigation water requirements are done while preparing national water master plans or irrigation projects. Useful information was obtained from a number of country studies available from AQUASTAT, but the information was based on many different approaches. For the purpose of this study the need was felt to develop a method of computing irrigation water requirements for the whole continent in a systematic way. In order to be able to do this at the scale of the continent, assumptions have to be made on the definition of areas to be considered homogeneous in terms of rainfall, potential evapotranspiration, cropping pattern, cropping intensity and irrigation efficiency.

The criteria used for the delineation of the irrigation cropping pattern zones were, in order of decreasing importance: distribution of irrigated crops, average rainfall trends and patterns, topographic gradients, presence of large river valleys, presence of extensive wetlands, population pressure, technological differences and crop calendar above and below the equator. The starting point was the type of irrigated crops currently grown in Africa. This resulted in 18 zones. From these zones, sub-zones showing a different cropping intensity or a different crop calendar were defied. Only the main crops currently grown, those occupying at least 85% of the irrigated area, were considered. Land occupation of the remaining 15 % by secondary crops was assigned to the main crops. An 'average' typical monthly crop calendar was assigned to each zone, based on work done by FAO's global information and early warning system, and on information from the reference library of FAO's agro-meteorology group, AQUASTAT and, for eastern Africa, from the IGADD crop production system zones inventory.

For each crop the actual cropping intensity was derived from national crop production and land use figures extracted from the FAO AGROSTAT and AQUASTAT databases. It ranges from 100 to 200%, according to the crop calendar. The cropping intensity to be used in this study of irrigation potential was generally estimated by increasing current values by 10 to 20%, but it was assumed that because of market limitations the current high intensity of vegetables in certain parts of the continent would not be found in the potential scenario. Therefore, intensities of cereal crops are higher in the potential scenario than in the actual situation.

This book aims to acquaint readers with what entails water technology management, especially as the challenges of water preservation.

— *Kenneth Johnson*

Irrigation Methods and Techniques

Methods

With agriculture responsible for the largest water usage in the United States and with irrigation dams being the most common type of water supply dam, it is important to examine the way this industry uses water and how conservation methods can be used to increase efficiencies and thus possibly decrease the need for dams. In addition to some of the alternative diversion techniques (described above) to supply water for irrigation, the U.S. EPA has compiled water-saving irrigation practices into three categories:

- Field Practices
- Management Strategies
- System Modifications.

When these practices are combined with the alternative diversion strategies above, the need for a diversion dam for irrigation could be eliminated in some circumstances.

Land has been tilled and terraced to better capture water (Lynn Betts, USDA Natural Resource Conservation Service)

Field Practices

Field practices are techniques focused on keeping water in the field, distributing it more efficiently, or achieving better soil moisture retention. These techniques are typically less expensive than management strategies or system modifications. When traditional field practices fall short of expectations and the management strategies and systems modifications discussed below are out of reach, the field practices of dryland farming and land retirement are another avenue to explore. Examples of field practices include:

The chiseling of extremely compacted soils;

- Furrow diking to prevent runoff;
- Land levelling for a more even water distribution
- Dryland farming; and
- Land retirement.

Farmers can develop land management practices that will decrease the demand on water supplies. More than half of land used for agriculture is still irrigated via a gravityflow system. This system uses soil borders, furrows, or ditches in order to allow gravity to distribute water across fields. Gravity flow irrigation methods can result in up to 50 percent water loss due to evaporation, inefficiencies in water delivery to the crop-root zone and runoff at the end of the field. The traditional gravity-fed system can be improved upon with the use of laser levelling or micro irrigation, though evaporation still leads to water loss. Laser levelling involves grading and precisely levelling the soil to eliminate any variation in the gradient and reduce slope of the field. This helps control the flow of the water and allows for more uniform soil saturation. Another method of preventing runoff is furrow diking. Furrow diking is the practice of building small temporary dikes across furrows to conserve water for crop production, which may also aid in preventing erosion.

If the above land management practices are not decreasing water use enough and the system modifications described below are too cost prohibitive or not an appropriate technique for a particular crop, farmers can also consider converting to dryland farming, switching to less water intensive crops, or land retirement. Farmers practicing dryland farming in arid regions use a variety of techniques and land management practices to minimize water loss and erosion. These techniques include coordinating seeding to the ideal soil moisture content, choosing crops more suited for arid conditions, and fallowing. Fallowing refers to a number of practices used for well over a century, such as plowing a field in late fall or early spring to clear weeds and increase soil moisture. Initial plowing breaks up the land and allows the soil to absorb more water. It also eliminates moisture-sucking weeds and creates ridges in the land that limit runoff and better capture moisture from snow. Fallowing can also involve choosing not to plant a certain field for one or more growing seasons.

Land retirement refers to a common policy of permanently or temporarily suspending farming on a particular acreage of land in exchange for financial incentives. One of the best-known land retirement programs is the U.S. Department of Agriculture's Conservation Reserve Program (CRP). Through CRP, farmers are paid annual rent per acre and an additional sum for providing land cover. While CRP has typically been utilized to control the agricultural market and keep prices and quantities stable, the added value of conserving land and water resources has been given more consideration in determining compensation for land retirement since the late 1990s. This type of financial incentive is common among land retirement programs.

Advantages: Practices such as chiseling, furrow diking, and land levelling allow the land to absorb water more efficiently and results in less waste. It is also one of the most inexpensive methods of agricultural water conservation discussed in this report. Depending on the amount of land in need of irrigating and the alternative chosen, it might be possible to remove an irrigation diversion dam, particularly if used in combination with one of the alternative diversion methods described above. Dryland farming and land retirement, also discussed above, have the most to offer in terms of water savings, simply because they call for the use of little to no water, and the potential for dam removal.

Disadvantages: While chiseling, furrow diking, and land levelling help prevent runoff and allow the land to retain more water, they still do not address the over watering that results from gravity-fed irrigation. Also, dryland farming and land retirement practices can seem akin to suggesting that farmers go out of business. Discussions centring on these alternatives should take current use and compensation into consideration. Also, dryland farming and land retirement practices are rarely, if ever, applied to the large agribusinesses that now dominate the industry.

Costs: As discussed above, furrowing and other land levelling practices are the least expensive irrigation alternatives discussed in this report. Actual project costs will vary depending on amount of acreage, topography of the land, and the region or country in which the farm is located. According to the 1998 Farm and Ranch Irrigation Survey, capital expenditures in the United States for farm improvements were $643 million for irrigation equipment and machinery, $138 million for construction and deepening of wells, $190 million for permanent storage and distribution systems, and $83 million for land clearing and levelling. In order for dryland farming and land retirement to be feasible for farmers, it often must be accompanied by financial incentives like conservation easements, which involves the transfer of development and/or land use rights to a government agency or non-profit providing tax benefits or direct payment for retirement of the land.

Management Strategies

Management strategies allow the irrigator to monitor soil and water conditions to ensure water is delivered in the most efficient manner possible. By collecting this information, farmers can make informed decisions about scheduling, the appropriate amount of water for a particular crop, and any system upgrades that may be needed. The methods include:

- Measuring rainfall;
- Determining soil moisture;
- Checking pumping plant efficiency; and
- Scheduling irrigation.

Farmers have to rely on a number of factors to monitor soil moisture, including temperature and humidity, solar radiation, crop growth stage, mulch, soil texture, percentage of organic matter, and rooting depth. A variety of tools for monitoring soil moisture, such as Time Domain Reflectivity (TDR) probes or tensiometers, are also available to farmers.

The government of Queensland in Australia has done an effective job of compiling a fact sheet on a variety of irrigation scheduling tools, including the associated pros, cons, and costs of each. Ensuring that pumping plants are running at their most efficient also guarantees that water is being delivered to the plant and not wasted. Efficiency can be checked by examining the volume of water pumped, the lift, and the amount of energy used. A pump in need of repair or adjustment can not only waste water but also cost money.

Advantages: The management strategies described above allow for the correct amount of moisture to be delivered to the plant. When combined with system upgrades like the ones discussed below, farmers can maximize the amount of water savings and the efficiency of their land. While this is not an automatic replacement for a dam, there could be an opportunity for removal or the ability to delay construction a new barrier, depending on the size of the diversion.

Disadvantages: Monitoring the water needs of crops in the most efficient manner possible requires technological upgrades that require an initial outlay of capital. In addition to the cost of implementing these system upgrades, there may be training required to integrate new computer systems and other technologies.

Costs: Depending on extensiveness of the system, costs can vary significantly for the management strategies discussed above. For example, the average price of a tensiometer ranges from $120 to $200, with the average field requiring a minimum of four stations containing two tensiometers each, while acprobe system containing probes, training, and software can run as much as $9,120.

The Department of Natural Resources, Energy and Mines in Queensland (DNREM), Australia has put together a comprehensive fact sheet that provides cost estimates (in Australian dollars) for a wide range of irrigation scheduling tools.

A centre pivot irrigation system with drop tubes (Tim McCabe, USDA Natural Resource Conservation Service)

System Modifications

System modifications, often the most expensive of the three categories, require making changes to an existing irrigation system or replacing an existing system with a new one. Typical system modifications that allow for the most efficient delivery of water are:

- Add drop tubes to a centre pivot system
- Retrofitting a well with a smaller pump.

Replacement irrigation systems include:

- Installing drip irrigation, microsprinklers, or solid set systems; or
- Constructing a tailwater recovery system.

Many farms still use inefficient irrigation techniques (e.g., travelling gun, centre pivot) that apply more water than crops require. Modern irrigation technology, such as drip irrigation, micro sprinklers and solid set systems can deliver water much closer to the actual plant and achieve much greater water efficiency. These irrigation tools are the most efficient in terms of delivering water to crops. They use the latest technologies to determine the exact amount of water a crop needs in order to grow and delivers the water directly to the plant. However, they often prove most efficient when used with vegetable and fruit tree crops and less so with dense grain crops.

Advantages: Because of the considerable amount of water used in agriculture, improving efficiency in this sector offers an opportunity to achieve significant reductions in water use. By using the latest technology available to maximize the efficient use of water, the need for some water diversions and dams can be eliminated.

Disadvantages: Switching to more efficient irrigation technologies is cost prohibitive for many farmers. Even though federal and state incentives exist, they are often inadequate to address the scope of the problem.

Costs: As mentioned above, initial costs of the latest irrigation technology can be quite high. For example, drip irrigation systems can cost on average $1,000 per acre to install necessary pumps and filters and $150 per acre per year for drip tubing.

A study done by Kansas State University Agricultural Experiment Station in October 2001 compared the costs of centre pivot, flood and drip irrigation systems. While the drip irrigation systems are typically more expensive to install, farmers are able to recoup some costs with savings from reduced water use.

Irrigation Scheduling

Irrigation scheduling is the process used by irrigation system managers to determine the correct frequency and duration of watering.

The following factors may be taken into consideration:

- Precipitation rate of the irrigation equipment-how quickly the water is applied, often expressed in inches or mm per hour.
- Distribution uniformity of the irrigation system-how uniformly the water is applied, expressed as a percentage, the higher the number, the more uniform.
- Soil infiltration rate-how quickly the water is absorbed by the soil, the rate of which also decreases as the soil becomes wetter, also often expressed in inches or mm per hour.
- Slope (topography) of the land being irrigated as this affects how quickly runoff occurs, often expressed as a percentage, i.e. distance of fall divided by 100 units of horizontal distance (1 ft of fall per 100 ft (30 m) would be 1%).
- Soil available water capacity, expressed in units of water per unit of soil, i.e. inches of water per foot of soil.
- Effective rooting depth of the plants to be watered, which affects how much water can be stored in the soil and made available to the plants.
- Current watering requirements of the plant (which may be estimated by calculating evapotranspiration, or ET), often expressed in inches per day.

- Amount of time in which water or labour may be available for irrigation.
- Amount of allowable moisture stress which may be placed on the plant. For high value vegetable crops, this may mean no allowable stress, while for a lawn some stress would be allowable, since the goal would not be to maximize production, but merely to keep the lawn green and healthy.
- Timing to take advantage of projected rainfall
- Timing to take advantage of favourable utility rates
- Timing to avoid interfering with other activities such as sporting events, holidays, lawn maintenance, or crop harvesting.

The goal in irrigation scheduling is to apply enough water to fully wet the plant's root zone while minimizing over watering and then allow the soil to dry out in between waterings, to allow air to enter the soil and encourage root development, but not so much that the plant is stressed beyond what is allowable.

In recent years, more sophisticated irrigation controllers have been developed that receive ET input from either a single on-site weather station or from a network of stations and automatically adjust the irrigation schedule accordingly. When properly set up and maintained, these controllers do tend to conserve water over conventional human scheduling as the program is updated at least daily. Other devices helpful in irrigation scheduling are rain sensors, which automatically shut off an irrigation system when it rains, and soil moisture sensing devices such as capacitance sensors, tensiometers and gypsum blocks.

Choosing the most Advantageous Field for Drip Irrigation

Drip irrigation is gaining large amount of popularity due to the wonderful amount of benefits that is being provided by this irrigation technique. Nowadays there is more and more number of farmers who are implementing drip irrigation method which gives them the opportunity to automate the process of irrigation. In order to maximize the amount of benefits that are derived through drip irrigation it is necessary that the person possess adequate knowledge regarding the selection of proper field that will provide him best results. This article will be providing good deal of idea that will help in increasing the harvest of crops and also improve the quality of plants. The factor which influences the selection of field is technology, nature of crops, climatic conditions, labour inputs and many more.

Soil Selection in Fields

In drip irrigation there will constant spraying of water uniformly around the field. The soil type that is most suitable for drip irrigation is sandy soils. Surface irrigation cannot be applied for sandy soils since these soils have high penetration and very less storage capacity. Thus drip irrigation is the only best method that makes sure that proper level of water is supplied to plants without water wastage. It is also essential to note that clay or loamy soils should not be selected for fields that are being irrigated by drip irrigation since it will make soil slippery and water logged.

Climatic Conditions and Slope of Field

Drip irrigation can be greatly affected if the wind velocity in field region is on the higher side. When the wind velocity is higher then the throw of water will be disrupted and non uniform spread of water will be made. The effect of drip irrigation can be maximized in which the flow of wind is uniform and stable. Another important factor which influences field selection for drip irrigation is slope of field. This irrigation technique will provide more benefits if field is sloppy or non uniform since it will ensure proper path for flow of water. Best example for sloppy field is rice grown in terrace space. Surface irrigation cannot be applied in regions which contains slope.

Water Quality and Type of Crops

One of the important requirements for drip irrigation is quality of water which is present in field. It is essential to select field space in which water contains very less level of sediments or salt particles. Presence of higher level of sediments or salt particles will clog the sprinklers and decreases the efficiency of water spray. Since there is large level of investment which is being made for this irrigation technique only those fields which supports growth of higher value cash crops should be selected to ensure the farmer gets good returns for the money invested

Available Labour

Drip irrigation provides the possibility of reducing the labour requirement due to automation process. Fields where lower level of labours is available will be well benefited by adapting drip irrigation technique. It is essential to maintain sprinklers in good condition to maintain efficiency in operation.

Using Drip Farming in Winter

Agriculture has found drastic changes in past few years. The world has witnessed revolutionized approach in carrying out tasks in the field. The on-field agricultural practices have been greatly supported by inventions and discoveries that have hit the science world. Also usage of sophisticated tools and equipments has gathered substantial interests of farmers. This has empowered them to build up comprehensive farming with less effort generating huge dividends for the money invested. Importance of ambient conditions can be visualized with development of hybrid seasonal crops. The modern scientific aids to modern agriculture are also provided through researches carried out all over the world. Farming has now channelled its path towards thoughtful use of conditions with optimized usage of resources available. For instance, one can get huge throughput by planting right crops at right seasons. For instance during summers, one can engage cultivation of crops that favour higher temperature during their growth. This makes the overall process of farming refined totally conducive for people to march down to sustainable development.

Basic Irrigation Facilities

Irrigation plays a major role in the development of crops. Water can be utilized judiciously for the growth of the crop. There are various types of irrigation facilities available for the farmers present globally. Starting from ground water supply to supply from rivers. Researchers have also modularized the irrigation facilities to cater to ever-increasing demands. The use of sprinklers can be used to cover large area of field in short duration of time. This can also be by-passed by usage of overhead irrigation. Some of classical methods of irrigation include terracing and furrow irrigation. Furrow irrigation involves constant monitoring of farmers to regulate channeling of water to respective fields. Recent developments of science and technology have included drip irrigation to this long list.

Importance of Drip Irrigation

Drip irrigation involves the type of irrigation wherein the supply of water is confined to the root zones of each and every plant. This can greatly lessen the fritter of water. Also one can be sure of getting necessary supplied nutrition to the plant. Every other types of irrigation include surface evaporation of water. This can reduce moisture content that is required for the growth of plants. The pressure of water channel can send the water deep down into the earth at desirable spots near the root zone. Also by utilizing efficient computer module, one can monitor the flow of water and timing of watering to crops. This can be cashing for labour in many ways but the operating cost can fetch fruitful results for farmers. In recent years, sprinkler type of irrigation is replaced by unique type of drip irrigation referred as subsurface drip irrigation.

Usefulness of Drip Farming in Winter

Drip farming includes the laying down structural lines for channeling water to respective spots underneath the soil. This can also be aided by several hydraulic components such as pressure relief valves and hoses. These can have handsome impact from lowering of temperature during winters. The temperature changes can affect any physical element by changing the microscopic properties of materials used. Drip farming has major impact during the winter as it can get hoses and hydraulic accessories to get frozen with water that is capsulated within them. So to avoid mishaps, one can adopt popular winterizing technique. This involves expelling the water contained through special equipments that employs purging air to accomplish this act. Blown air would remove all contaminants and undesirable particles present in hydraulic lines. Electronic components associated with drip farming methods can be stored in secured place avoiding direct exposure to moisture. Regular checking is to be instituted for retaining cleaner fluid lines over the field area during winter season.

Drip Irrigation Fittings

Agriculture is a vast field. Every human has to consume food in order to live. These food items are got only from the hardship of a farmer. As the technology improves many modern methods have evolved to sophisticate all including the farmers. Most of these focused only on reduction of burden on the farmer and not on the improvement of farming. One step was taken to provide optimized resource utilization in the agriculture field. The new invention and the one which takes care of this optimizing step is the drip irrigation process. Many are not aware of the drip irrigation process. Some of them are stubborn and are not ready to go in for another process. They feel safe and secure with the conventional process.

Why we have to go in for Drip Irrigation?

As we all no water scarcity is the most faced problem around the globe. Not sufficient amount of water is present with many villages and their crops get spoiled due to it. So saving of water is a must for all of us. This can be easily achieved by the use of drip irrigation process instead of the conventional process. The wastage of water can be drastically reduced when drip irrigation is followed. The working principle of drip irrigation process is very simple and uses small openings in the hose to water the plant. The holes are so small that only drops of water emerge from these holes. So the water usage rate is so minimum and this is even smaller than the water intake rate of the plants. So the wastage of water is minimal and the utilization of water is optimum. The type of hoses used for this process is the soaker hose. This hose has minute holes of 1/6 to 1/4 inch diameter. And this is also has unique capability to work both on the surface and under the surface. The most important thing that is to be mentioned here is about the fittings.

Fittings in the Drip Irrigation Process

Water leakage and water loss in the fitting is the major loss in the case of any hose usage. The wear and tear is mostly felt at the fitting region. This can be reduced drastically if the special fittings are provided. The fittings used in the drip irrigation process is one of its kind. The fittings are more precisely manufactured and they form a tight fit. The clearance in between the threads are so minimum. The material these fittings are made up of is the brass. In addition to it hoses also posses this capability to fit perfectly with any pipes.

Drip Irrigation Timer

Some of us would have come across the word drip irrigation. We all know irrigation is a process of providing water to the plants, but we don't know what is drip irrigation. Drip irrigation is nothing but, providing water to the specific plant at specific place at specific time. The conventional process of irrigation satisfies these thing, then why is that we should go in for a new process of irrigation. The answer to this question is very simple.

Why Drip Irrigation is Needed ?

The need for drip irrigation is due to the difference or advantage it has over the conventional process. The drip irrigation saves water while the others waste them. Water is a big problem faced by many of the countries. So saving water is very essential to all of mankind. The process of drip irrigation is very simple and its application is also very easy. The working principle involved is that the plants are provided water with minute holes in the hoses. These hoses are known as soaker hoses. The soaker hose works by the principle of capillary outlets. There are very small holes in the range of 1/6 to 1/4 of an inch. These holes form the outlet or source of water supply to the plants. The other great advantage in these hoses are that, it can fit with any type of fittings and pipes. This advantage is got by the hose due to the fact that it poses brass fittings at the ends. The higher end hoses even poses greater advantages than the lower end hoses. They are manufactured precisely to meet the standards. The hose can be placed above the soil or as in some cases buried in the soil. This is how people grow plants in dry places. They are highly advantageous and capable of providing plants in any soil conditions. The other thing is that, it saves water by reducing the sprinkling and showering.

Drip Irrigation Timing

The other important thing that is so far not mentioned about the drip irrigation is that it can be integrated with mechatronics system. The water is provided to the plants based upon the actuation of valves. The valve opening and closing time can be automated using mechatronics system. So the water is provided to the plants at proper intervals and proper amount of water can be given to the plants. The advantage of the drip irrigation is doubled on application of the timing system in the process.

Drip Irrigation Soaker Hose

Irrigation is a common terminology to all of us. But the real meaning of it, is not known by many. Irrigation is a process of watering the plants artificially. This is done because, sufficient amount of water may not be present in that area or the season may not be correct for the cultivation. There are many artificial ways of irrigation possible. The normal way of irrigation is the way in which the water is sent to the plants via a channel or tunnel and sprayed at the end. But the major problem faced in this process is that, the amount of water wasted in this is too high. The problem solver that originated due to tiresome effort of many scientist is the Drip irrigation process.

What is Drip Irrigation Process?

Drip irrigation is a new and advanced process, that is not known by many. The way it differentiates from the other process is by, making the water directly available to the most needed part, rather than spraying it or converging a large amount through channels. This is achieved by making small openings in a pipe and focusing the water supply to the plants. The advantage associated with it is that it can save water to a large extent. The other thing about it is that it can even make a dry land into a fertile land. It is environmental friendly and to add to its advantages it can even challenge the nature itself. This was developed due to many hours hard work of many scientist. It is environmental friendly and accepted world wide for its capability. The limiting factor for this process, is the efficiency of the soaker hose.

What is Soaker hose and why it is Needed ?

All would have come across what a irrigation process is and what are the advantages associated with it. The item required to incorporate this method on the field is a soaker hose. The soaker hose is a item which resembles the structure of a normal garden hose. The difference it has over the garden hose is that, it has minute holes throughout its external surface.

This forms the outlet for water to escape and irrigate the plants at particular places. The holes are so small that each hole is of the diameter of one by sixth to one by fourth of an inch. This is the working principle of a soaker hose. The unique advantage of the soaker hose over the other hoses is that it can fit with and pipes and fittings.

Maintenance of Drip Irrigation Systems

Agricultural practices have refined in last few years. People have broader idea about getting fruitful results out of their investment on the lands. One can ensure effective growth of plant only if the plant or whole plantation is administered with good irrigation facilities. These facilities can be helpful in providing useful gains for the farmers. By far technological advancements have gathered minds of farmers to render useful ideas into planned actions. These technological advancements can be adjudged with useful application in day-to-day life.

Importance of Irrigation System

Irrigation plays a vital role in the modern day agriculture. The root tips are in search of water when they travel down inside the ground. These can be ably supported by irrigation administered by the cultivators. The irrigation methods are of varied types that are available for cultivators. Starting from drip irrigation to centre pivot irrigation. Some of irrigation methods take care of all the necessary topological features that prevail in particular land. The contour of surface changes from place to place. It is quite evident that it can affect watering scheme adopted by the farmers. But this would be justified by the amount of water needed. Some crops would need large supply of water to be supplied to root tips.

Irrigation Facilities Promised by Drip Irrigation

Drip irrigation is one of the popular methods adopted by the farmers all over the world. This method could bestow useful irrigation to the crops that are planted. Drip irrigation is preferred in the gardens. The gardens that are small can be well sheltered under effective watering through drip irrigation scheme. Also some watering scheme require some additional components such as tubing that can add to the cost of establishment of watering facility. This can be cashing for the farmers initially but with effective maintenance it can fetch huge profits on long run. Drip irrigation also ensures correct level of moisture that is to be maintained around the root tips. Any variation in this can cause adverse effects on the plant.

Drip Irrigations Systems – Components

Drip irrigation involves use of valuable wit to design the overall plan of the system. The system includes tubing and piping needs to be satisfied before going for commissioning of irrigation system. There are several components that form integral part of drip irrigation systems. Also one needs to ensure proper design of irrigation systems can be employed to facilitate moisture monitoring. The shoot system of the plant needs supply of water and other nutrients through root. The watering of plants at optimum level ensures proper supply of nourishments for the development of crop into sizeable yielding. Special simulation software is available to map all the essential failure nodes that can lead to drastic changes in the irrigation channel. Hence it is advisable to use that software to clear out all the loopholes present in irrigation channel.

Effective Maintenance of Irrigation Systems

Maintenance of irrigation system largely depends on the design of irrigation system. The maintenance of the drip irrigation system involves series of actions that can ensure proper working of overall system. Subsurface drip watering schemes are widely used by the farmers pertaining to different parts of world to irrigate the lawns. These can be effectively monitored for pressure inside the channel. This pressure if not checked periodically can institute massive technical failures for the entire irrigation system. People can seek expert advice to have proper maintenance. The tubing can be checked for correctness periodically to eliminate water leakage. The inlet and outlet nozzle can be cleared off from dust and mud in order to have proper entry and exit of water inside the whole watering system.

Types of Drip Irrigation Valve

If you want to irrigate your farms with a drip irrigation system, having the right drip irrigation valves properly installed is a must.

There are various types of drip irrigation valve you can find in the market. Most drip irrigation systems need at least two different purposive-types: an emergency shut off one and a controlled one.

Emergency Shut Off Valves

Emergency shut off valves are drip irrigation valves to be installed in the area where you tap in for your drip irrigation system. It is the closest point to water source. Having these valves installed is crucial. Without them, you will have to shut off the water flow to your entire house should an irrigation breakdown occurs. Without them, you will also have to work on the mainline or irrigation valves.

The most commonly used for this purpose are gate valves. They are quite affordable though also tend to wear out quickly and then start leaking. Better recommended are ball valves, disk valves and butterfly valves. They are more reasonably pricey than gate valves since they are much more reliable and last several times longer.

Zone Control Valves

Zone control valves are drip irrigation valves that turn on and off water to the drip tubes (or hoses). Often these valves are automatically turned on and off by an irrigation controller/timer. For a small drip irrigation system you may need only one zone control valve. While for bigger drip irrigation systems several more may be required. There are two basic types of zone control valve: standard globe valve and anti-siphon valve.

Standard globe valves can be found in almost any size. They are often installed below the ground, kept inside a box or a vault. Since a standard globe valve does not incorporate a backflow preventer, you must provide one on your own.

Anti-siphon valves can be found in 20mm (3/4 inch) and 25mm (1 inch) sizes. They are most recommended for home gardeners since they incorporate a backflow preventer, thus saving a considerable amount of money. The valves must be installed above the ground, at least 150mm (6 inch) higher than the highest drip emitter. This requirement may be a problem for certain areas, but you can always install anti-siphon valves on top of trellis or at the top of the slope (if you have a garden with slopes).

Indexing Valves

Indexing valves are a single valve unit purposely installed to control several valve zones. Usually available in models with or without a built-in anti-siphon device, it requires a special controller to operate. It has a water inlet and several water outlets. When the control unit sends the first signal it opens the first water outlet. At the next signal it switches to the second water outlet. And then to the third, fourth and so on till it gets back to the first water outlet, at which point it shuts off.

Indexing valves must be installed above the ground, at least 150mm (6 inch) higher than the highest drip emitter. It has never been too popular and is generally available in certain regions where a nearby manufacturer has promoted them.

Drip Irrigation Hoses Available in the Market

Drip irrigation hoses, also known as drip irrigation tubing, are hoses specifically designed to carry water and drip it through tiny holes attached to them; the tiny holes are called "emitters". They drip water slowly into the soil exactly at the plant root zone where it is needed. This way moisture levels are kept optimal, improving plants productivity and quality.

Drip irrigation hoses are made of polythene. They come in various types and diameters, accordingly to your needs. The length of a single drip irrigation hose should not over 200 feet from the point where water enters it.

You need to stake the hoses to keep them from moving. More importantly, never bury the hoses and their emitters even if they are designed to be. Otherwise, you will need to spend more time and energy to overcome clogging and rodent damage.

When choosing hoses for your drip irrigation system, you have to keep in mind these important factors: size, pressure rating, weight, length and chemical compatibility.

Some Drip Irrigation Hoses Available in the Market

Drip Irrigation Hose.580

It comes in various lengths: 25', 50', 100' and 500'.

Using.580 compression fittings, it operates at pressure rate 10 to 60 PSI and has a maximum flow rate 180 GPH. It is best if you operate it at pressure rate 25 PSI.

Drip Irrigation Hose.700

It comes in various lengths: 50', 100' and 500'.

Using.700 compression fittings, it operates at pressure rate 10 to 60 PSI and has a maximum flow rate at 240 GPH. It is best if you operate it at pressure rate 25 PSI.

Drip Irrigation Hose.820

It comes in various lengths: 100', 250' and 500'

Using.820 compression fittings, it operates at pressure rate 10 to 60 PSI and has a maximum flow rate at 380 GPH. It is best if you operate it at pressure rate 25 PSI.

1/43 Micro-Drip Irrigation Hose

It comes in various lengths: 25', 50', 100' and 500'. Used with 1/4" barbed fittings, it is used for extending drippers and micro-sprinklers from main line or as the primary line in a small drip irrigation system. It can be put above or below the ground. Use this hose to distribute water from main line to drippers, misters and low volume sprinklers.

1/43 Laser Drilled Drip Irrigation Hose

It comes in various lengths: 50' and 100'. Holes are laser drilled into the hose against the flow of water. Used with 1/43 barbed fittings, it is to be put within 12" or 6" space.

Advantages of a Subsurface Drip Irrigation System

Drip irrigation systems are mostly installed permanently or semi-permanently above the ground. When applied to irrigate trees and vine crops, they promote no problem. However, when applied to irrigate field and vegetable crops, certain problems occur, mostly because the drip irrigation systems have to be installed and retrieved annually. The annual handling is harmful to the life time span of the system, even under the best handling conditions. Hence the subsurface drip irrigation system emerges.

The depth installment of subsurface drip irrigation system is determined by the soil type and the crops planted. An efficient installation has water moving at the depth of 4 to 30 inches beneath the surface, forming a continuous watered area along the plants row. Frequent irrigation cycles, up to several times every day, maximize capillary action and minimize water surfacing.

Four subsurface drip irrigation systems have been designed, installed and successfully produce better yields by the Water Management Research laboratory of the USDA in Fresno. The oldest one was installed in 1981 and used for three years to irrigate tomatoes. During those years, they also successfully produced twice as much broccoli, at the same farm. Subsurface drip irrigation system allows the precise application of water, nutrients and other agricultural chemicals directly to the plants root zones. It allows the optimization of growing environment, which leads to higher quality and quantity crop yields.

With the same amount of water, subsurface drip irrigation system waters 46% larger volume of soil than any conventional (surface) drip irrigation system. It decreases the soil saturation, which not only leaves room for more air, but also improves water capillary movement and decreases water loss due to deep percolation. The top 15 – 20 cm of soil remains dry, therefore evaporation of water at the top of the soil may decrease and less salt will accumulate at the surface. Permanent installation of subsurface drip irrigation system also provides considerable labour saving since irrigation can be applied while the whole equipment stays in the farm.

Any soil surface crusts which usually lead to infiltration problems are no longer trouble makers; they are bypassed if this subsurface drip irrigation system is installed. Other farming applications required to support the conventional drip irrigation system is no longer required, too. Since subsurface drip irrigation system is installed below the ground, all components are not exposed to sun light, constant wetting, weather changes and other environmental changes. It is expected to last longer than the conventional drip irrigation system.

Drip Irrigation Greenhouses

Greenhouses type of plantation has gained massive support from the people all around. They need to have better growth under specified condition. Also the growth of plants can be termed healthy due to procurement of essential conditions that are for the uninterrupted growth of plants. The harmful radiations can be stopped from approaching the plants productive systems. Conducive conditions are reached at the greenhouses instituting better growth of plants.

Essentials for Drip Irrigation

Essentials for any type of irrigation system are water supply. The source of water can determine the type of irrigation employed. The drip irrigation involves effective management of the water through proper channel. The channeling of water can be established by rational drip irrigation design. Also the moisture level at the region beneath the soil surface to be regulated periodically with drip irrigation system

Drip irrigations are essential in area wherein there is scarcity in water.. Greenhouses are generally associated planned agriculture carried out at the expense of quite substantial investment. The effectiveness of the greenhouse is greatly determined by the quality of the plants grown so avoid any kind of pest or weed attack to the greenhouse plants, one needs to adopt for irrigation. It eliminates the opportunity for the growth of other types of plants like weeds taking nutrients from the ground surface.

Design your own Greenhouse Drip Irrigation

Drip irrigation can be customized based on the type of the field that is to be irrigated. The irrigation system involves the use of tubing, valves, pressure regulators and other plumbing accessories to get planned irrigation method. Pressure flow regulators to a maximum value of about 30 psi can regulate the operating pressure of tubing. The slope of field, soil features and water source can be carefully studied before validating a design. The water requirement for a particular variety of plant can determine particular type of accessory like emitter to be used.

Drip Irrigation Vegetable Gardens

Conventional irrigation system is confined with number of disadvantages to be associated with. The development of the facilities in agriculture has prompted people to adopt drip irrigation for carrying out all agricultural practices.

Drip Irrigation for Vegetable Gardens

Vegetable gardens are irrigated by drip irrigation in modern agricultural practice.. Gardens employ different kinds of plants to be grown under one shelter. This makes the work of gardener to be difficult in maintaining the garden. So it is advisable to use drip irrigation technique to have programmed watering of plants. This enables proper maintenance of moisture level in the ground. Since vegetable plants require different watering for obtaining better growth of plants.

Tips for Getting Efficient Drip Irrigation for Vegetable Gardens

It is essential to maintain adequate level of pressure in the tubing of drip irrigation. The micro irrigation technique can aid planned watering of younger plants. The watering of large and grown plants can be done using different types of emitters. Overall usage of water can be optimized with some practical testing of the drip irrigation system after installation. Vegetable plants like carrot can require only adequate amount of moisture to be maintained through out its growth cycle. So it is advisable for the designers to use multiple emitters and bubblers to carryout watering of those plants. Stream bubblers can be used for watering plants like tomato. The constant moisture can be preserved in the vegetable garden by drip irrigation technique.

Advantages of Drip Irrigation in Vegetable Gardens

Vegetable gardens with drip irrigation involve splitters used to channel the water for different plants. These can reduce the overall wastage of water yielding to effective water conservation. Drip irrigation provides random type watering enabling elimination of the weeds in vegetable garden. Vegetable gardens are prevented from soil erosion by using modular drip irrigation system.

Drip Irrigation Pressure Regulator

Drip irrigation is popular among all people residing all over the world. The utility of drip irrigation is characterized by the presence of certain implied conditions to be satisfied. These technical requirements determine the overall activity of the drip irrigation system.

Technical Requirements

The pressure is the main ingredient that is to be maintained by tubing with adequate level of supply of water. The tubing should withstand a pressure of about 2 bars. The tubing is to be maintained sufficiently deep in the ground. The ground can be dug to a depth that can be easily approached by the root tips of plants. Pressure regulation can be done to minimize or maximize the pressure that is maintained. Also flow valves can control the intensity of water. The ideal flow that is used is about 0.2l/s for standard dimension of tubing. The tubing that is generally employed in the drip irrigation is about 24mm diameter.

Pressure Regulator-Types

Pressure regulator is of different types. Varied range of pressure regulators can be used for the effective control of pressure in the tubing in different parts of the drip irrigation systems. They are numerous types and differ based on the design specifications namely—

- Low flow pressure regulators. The design of low flow regulators is quite unique. Also the design can regulate pressure based on the utility. A flow rate of about 0.1 to 1GPM can be achieved with low flow pressure regulators.

- Medium flow pressure regulators. Medium flow regulators are utilized for the gardens that require the flow rate of range 2-20 GPM to be achieved. With this type of flow pressure regulators one can limit the pressure based on the inlet pressure applied.
- High flow pressure regulators. Flow rate can be limited to about 32 GPM with this type of high flow pressure regulators. The external pressure can vary from 10 psi till 50 psi.

Traditional Irrigation Fails Themoney Test

OUT OF the total 303.42 million ha of land that make up India, 136.18 million ha were categorised in 1987-88 as net sown area and of this, only 43.05 ha receives irrigation. The sources of irrigation, according to official records, are canals, tubewells, tanks, wells and other sources, a classification adapted by the British for the convenience of revenue categorisation.

Nirmal Sengupta traces the evolution, decline, current status and future prospects of what has been categorised as tanks and 'other sources', which broadly, though not exhaustively, include wells, irrigation channels and lift irrigation schemes. These provide water for about 6.24 million ha. The book is fascinating both because of the subject with which it deals — user-friendly (participatory) irrigation systems — and the manner in which it has been treated.

Narrating the schism in the usage of the terms traditional and modern, Sengupta desists from associating traditional with participation or user-friendliness though he makes no bones about the connection.

Old is Effective

He kicks off with the argument that "traditional" irrigation systems cannot be done away with because the terrain of our agricultural lands does not permit modern canals and tubewells to be effective in even one-fifth of the country. Therefore, future increase in irrigation capacity and even sustenance of current levels will depend on the expansion and maintenance of existing systems. The fact that "traditional" systems exist even today is a testimony to the sustainability, eco-viability and efficiency of these designs.

Traditional techniques such as canal irrigation, which have been studied and developed by engineers of this era, have come to be labelled as "modern" systems. The process of selection has been based not on whether the system is traditional or modern, nor on the large-small divide; the dividing line is between people's participatory systems and centralised bureaucratic systems.

Volte-face

However, when it comes to the economic viability of these systems, Sengupta does a volte-face. He argues that financially, these "traditional" systems do not pass the test. The modern techniques, which were hitherto argued against so forcefully, suddenly become acceptable and the lacunae of them suffering from bureaucratic control can, it seems, be overcome by putting more effort into user participation while operating the system.

At the end of it, the reader is left slightly confused. Is the book arguing for the current thrust in irrigation planning to be changed from mega-projects, which by necessity of design have to be centrally-controlled and unwieldy, to specific, user-need based projects, or is there a plea for smaller designs serving a smaller localised command and allowing people's participation? This is where Sengupta should have, but did not, draw a link between design, planning, control, management and social structure, and given his work a holistic perspective. Any water storage system or run-of-the-river facility has integrated links with the people it affects in more ways than one.

Any analysis of irrigation systems cannot be bereft of an understanding of the complex linkages on each of the above counts. Sengupta has, however, made no such attempt. Modern canal systems have linked with them issues such as displacement, resettlement and rehabilitation, changes in social and economic organisation in the local populace due to canal irrigation, agricultural practices and sustainability of resource use, which no sensitive study of irrigation can afford to bypass.

While initially Sengupta does talk about the tremendous sustainability of user-friendly designs, he completely ignores this question while analysing the cost-benefit aspects of modern systems, understandably using a very narrow definition of costs, benefits and efficiency.

However, the problems notwithstanding, any serious researcher on Indian irrigation systems would necessarily have to add this book to his reading list.

Modern Irrigation Techniques

In order to feed very large numbers of people, modern farming has become very efficiency-conscious. Every possible way of improving crop yields is pursued. Very large, complex, expensive equipment is now virtually required, in order for a farm to be profitable. This also means that extremely large farms must be used, to amortize the very high cost of such equipment. It also means that massive use of pesticides and herbicides is necessary to process such huge areas regarding insects and weeds. More to the point here, phenomenal amounts of water is needed to permit all those plants to grow at optimal rates.

Natural soil moisture and natural rain used to be the sources that were relied on. Modern farming cannot rely on the natural variations of such things. Irrigation is central to all modern farming.

Considering how cost-conscious and efficiency-conscious modern farming has become, it amazes me at how poorly irrigation is used! Farmers tend to operate their irrigation pumps while they are awake, during the daytime. That seems to make sense, for the possibility of any mechanical malfunction. But for several reasons, it is very poor use of that resource.

- First, during hot, sunny summer days, a significant amount of the sprayed water evaporates in the warm air between the time it leaves the spray heads and gets to the plants.
- Second, once water lands on the leaves of the plants, the bright sun and hot temperatures causes even much more evaporation of that water, before it can be of any use to the plants.
- Third, due to the surfaces of the leaves, many of the drops of irrigation water rest on the leaves as relatively spherical drops and not as a uniform wetting of the surfaces. This is especially true when a plant first begins to be sprayed. Unfortunately, one effect of this droplet shape is to optically focus the sunlight, very much as a glass lens would do. In the same way that a glass magnifying lens concentrates sunlight so much that it can quickly burn a piece of paper, these droplets can cause very small areas of the leaf surfaces to receive too much sunlight! Cell damage can result, even though later droplets of water are likely to cool it down to ameliorate this sunburn effect.

Regarding this third issue, many homeowners already know that it is a bad idea to water their lawns in the middle of a very hot, sunny, summer day, because the lawn can become "burned". Why don't the technological modern farmers know that?

An obvious solution to this exists! If irrigation was begun in the early evening, and potentially continued to a little after sunrise, fully twelve hours of watering is practical each day. FAR less water would be required, due to the much lower evaporation that occurs during the cooler, sunless nights. The plant leaves would similarly not have to endure localized focused areas of sunlight and the possible damage that can result from that.

The benefits from this simple change can be significant. First, from a practical view, irrigation pumps would need to run fewer hours to provide equivalent usable moisture for the plants, which would lessen equipment maintenance costs and the cost of the electricity to run those very large irrigation pumps. Second, if less water is needed to fulfil the needs of the plants, less water would be removed from the underground aquifers.

On this last matter, there seems to have long been an incorrect assumption that the amount of available underground water is unlimited. It is not. In the American Midwest, there are many thousands of cities and towns that drilled wells into a large aquifer, whose water seems to have fallen in the Colorado region. This water flows horizontally rather slowly, and it apparently takes hundreds or thousands of years to seep those many miles across the country. Since there are now thousands of agricultural irrigation wells that also draw water from this same aquifer, a problem is developing.

In recent years, it has been becoming obvious that aquifer is NOT of infinite capacity. Due to all these municipal and irrigation wells drawing so much water out, the level of the water in the aquifer has been dropping, and rather rapidly! Many wells that have operated properly for many decades have become "dry". The solution is now to have those wells drilled deeper, and many cities and towns and homes have had to do just that in recent years. But the pattern is alarming. If wasteful and extravagant use of fresh water continues, there will some day be a time when, no matter HOW deep those wells will be drilled, there will be no available water in that aquifer. Considering that water to replenish it may take hundreds or thousands of years to seep through the many miles of rocks to get there, a VERY serious problems seems likely to exist!

Now, no one knows how much that aquifer initially had, or to what percent it has already been depleted. There is not even very good data on the collective total of water that is being removed from it! Each farm considers the water under it to be "personal property" which can be used in any ways and in any amounts desired! Various States attempt to keep track of Municipal water use, but the data on irrigation consumption is very sketchy.

In any case, we can hope that we have only depleted 10% of that aquifer so far, which would imply that the millions of people who live in the Midwest can expect to have supplies of fresh water for at least a couple hundred more years, before the aquifer will be so depleted that Municipal water will no longer be available. But what if the massive consumption of water in the past 30 years has 50% depleted it? That would mean that crisis could occur in only another 30 years! Many people living now would be faced with a total lack of available fresh water. Catching rainwater might enable survival, but rainwater contains many contaminants that are picked up from the air as the droplets fall from the clouds, so it is not really a "safe" source of drinking water.

If, when, this crisis happens, there will really be NO solution! All of those farms that REQUIRE massive amounts of irrigation will no longer be able to operate. The farmers will likely just haul their equipment to Brazil or some other country where available farming land will still have available irrigation water. But the homeowners and the business owners and the towns, what will they be able to do? How could they ever live and work, if fresh water would not be available for hundreds of years?

I realize that, as long as politicians do not see a looming crisis on this issue, that nothing will be done. And even if new laws would be passed, how could anyone monitor the operation of wells on the PRIVATE LAND of the many farms? Considering that farmers know that they MUST use irrigation water in order to be profitable, is there even the remotest chance that they would just stop?

Can Irrigation be Sustainable

Globally about 10 Mha of agricultural land are lost annually due to salinisation of which about 1.5 Mha is in irrigated areas. While some climate and management aspects are common to semi-arid regions the detailed mechanisms and options to secure ecological sustainability and economic viability may vary considerably from case to case. This paper applies a whole of system water balance to compare irrigation in three semi-arid regions suffering from similar sustainability issues: Rechna Doab (RD)-Pakistan, the Liuyuankou Irrigation System (LIS) – China and Murrumbidgee Irrigation Area (MIA)-Australia. Soil salinity, lack of adequate water resources and groundwater management are major issues in these areas. The MIA and LIS irrigation systems also suffer from soil salinity and low water use efficiency issues. These similarities occur in spite of very different climatic and underlying hydrogeological conditions. The key data used to compare these different regions are climate and soils, available water resources and their use, as well as components of the water balance. In addition, the history of water resource development in these areas is examined to understand how salinity problems emerge in semi-arid regions and the consequences for production.

Based on the efficiency parameters and the definitions of sustainability, approaches are explored to solve common environmental problems while maintaining economic viability and environmental sustainability for irrigation systems.

Media Summary

It is possible to maintain the productive function of any irrigation area by providing adequate drainage and salt export facilities that may, however require high energy and capital investments. The most cost-effective option for sustainable irrigation is to increase water use efficiency and minimize negative impacts on the environment. There is a need to radically rethink sustainability of food production, rational pricing and sharing of water and commodities to justify the investment required to maintain and enhance ecosystem function within irrigated catchments.

Just 20 % of the world's croplands are irrigated but they produce 40 % of the global harvest which means that irrigation more than doubles land productivity (FAO, 2003). In developing countries irrigation improves economic returns and can boost production by up to 400%. On the other hand, irrigation can have unwanted environmental consequences. About one-third of the world's irrigated lands have reduced productivity as a consequence of poorly managed irrigation that has caused water logging and salinity (FAO, 1998).

Irrigation has been important for agricultural production in Mesopotamia (parts of present day Iraq and Iran) for 6000 years. The region has low rainfall and is supplied with surface water by two major rivers, the Tigris and the Euphrates.

The plains of Mesopotamia have always had problems with poor drainage of soils, drought, catastrophic flooding, silting, and soil salinity. Although Mesopotamia is very flat, the bed of the Euphrates is higher than that of the Tigris; in fact, floods of the Euphrates sometimes found their way across country into the Tigris. Engineers took advantage of this gradient as soon as irrigation schemes became large enough, by using the Euphrates water as the supply and the Tigris channel as a drain.

The main engineering problems for earlier civilisations were water storage, flood control and maintenance of canals. The salinity problem was more subtle, not fully appreciated, and could not be overcome by the engineering available at the time. It was difficult to drain water from fields, and there was always a tendency for salt to accumulate in the soil.

The problems of irrigated agriculture in Mesopotamia can be summarised as:

- *Silting of canals:* silt built up quickly in the canal beds, threatening to block them
- *Soil salinity*: recorded evidence around 2000 BC, 1100 BC, and after 1200 AD
- *Water politics* arising from tension between upstream and downstream users. In Sumeria, the city of Lagash was far downstream in the Euphrates canal system. The governor of Lagash apparently decided that he would cut a canal to tap Tigris water rather than rely on water from the Euphrates, but the addition of poor-quality water from the Tigris led to rapid salinization of the soil.
- *Over exploitation of resources*: After the wave of Moslem expansion overtook Mesopotamia, the Abassid Caliphate was based in Baghdad from 762 AD until its demise in 1258. Existing irrigation schemes were renovated and greatly extended in very large projects. Abassid engineers drew water from the Euphrates at five separate points, and led it in parallel canals across the plains, watering a huge area south of Baghdad. This system provided the basis for the enormously rich culture of Baghdad, which is still remembered in legend (Scheherezade, the Caliph of Baghdad, and the Arabian Nights) as well as history. But the scheme required a high level of physical maintenance, and there was increasing salinisation in the south.
- *Institutional failure*: As the central government began to fail in the 12th century (mostly from overspending), the canals became silt-choked, the irrigation system deteriorated, and the lands became more salinised. The deathblow to the system was aided by nature: massive floods about 1200 AD shifted the courses of both the Tigris and the Euphrates, cutting off most of the water supply to the Nahrwan Canal and wrecking the whole system. The Abbasids were too weak (or bankrupt) by

then to institute repairs, and the agricultural system collapsed. By the time the Mongols under Hulagu devastated Iraq and Baghdad in 1258 AD, they conquered a society that occupied wasteland. Iraq has remained a desert for more than 600 years.

Key Challenges to Irrigated Agriculture

About 50% of the total developed fresh water resources of Asia are devoted to growing rice (Barker et al., 2001). In Asia, current estimates show that by 2025, 17 million hectares (Mha) of the irrigated rice area may experience "physical water scarcity" and 22 Mha "economic water scarcity". It is projected that global rice consumption in 2020 will increase by 35% from the levels of 1995, whereas water availability for agriculture over this period is expected to fall from 72 to 62% globally and from 87 to 73% in developing countries. Increasing water scarcity threatens the sustainability of irrigated agriculture and hence food security and the livelihoods of rice producers and consumers.

Increasing competition from domestic and industrial uses has further compounded the problem of water scarcity. The demand for freshwater for industrial and domestic urban needs is growing rapidly throughout Asia. Less water will be available for agriculture and for rice, the crop that consumes the largest amount of freshwater. In some areas, water scarcity is already a major problem and a serious limit to agricultural development. Farmers are under pressure to grow more "crop per drop". Therefore there is an urgent need to find ways to "grow more rice with less water", to achieve this, efficient and appropriate irrigation technologies are needed.

Some of the challenges faced by present day irrigated agriculture are similar to ancient Mesopotamia and are summarised below:

- *More efficient use of inputs* (water, fertilizer, pesticides and labour) aimed at reducing negative impacts on the environment and to reduce production costs.
- *Soil salinity*: Matching landscape capability with irrigation systems, there is a need to manage salinity hazard by matching landscape capability when making decisions on new locations for irrigation development or on-farm field suitability.
- *Minimising environmental impacts*: Management of extraction impacts by quantifying both positive and negative externalities of different irrigation areas and sectors by evaluation, auditing and benchmarking in the irrigation industry. Management of

negative environmental impacts, such as methane and nitrous oxide emission, salinity, water pollution (abuse of pesticides), algal blooms etc., especially in intensive crop production systems.

- *Balancing consumptive and environmental demands*: Balancing irrigation and environmental flow demands through real savings due to improved distribution and on-farm water use efficiency and alternative cropping options.
- *Maintaining and enhancing quality* of drainage water and minimising impacts on rivers and ecosystems.
- *Institutional robustness*: Avoid a repeat of Mesopotamia in terms of institutional and ecosystem failures through better definition of property rights.
- *The Challenge:* to improve water use efficiency in rice-based irrigation systems of Australia and Asia

Three Regional Irrigation Perspectives in Rice-Based Systems

In 2000, the world's rice production was about 600 million tones Mt), 91 of which was produced in Asia. China, India, Indonesia, Bangladesh, Vietnam, Thailand, Myanmar, Philippines, Japan, and Brazil are the top 10 rice-producing countries (FAO, 2000).

Australia

In Australia, the rice crop is mainly grown on the Riverine Plain in New South Wales by utilizing surface water supply from the Murray and Murrumbidgee rivers and pumping from the Murray Aquifer System. Water is mainly supplied to the crop through a channel network serviced by irrigation companies or is pumped by the farmers directly from the rivers and creeks. According to the Australian Bureau of Statistics, gross water supplies in Australia to the rural sector (mainly used for irrigation) increased from 12.7 BCM (billion cubic metres) in 1983–84 to 15.8 BCM in 1995–96. Irrigated areas increased by 3.7% per year from 1.627 Mha in 1983–84 to 2.332 Mha in 1993–94.

Water use is currently 72% of the total water used for Irrigation in Australia. Surface water diversions in the Murray Darling systems are limited to a total of 11.6 BCM out of which around 1.6 BCM are used in rice production. All operations are mechanized. From 1.2 to 1.6 Mt of paddy rice are produced per year on 2300 farms. In 2001, 1.7 Mt of rice were harvested from an area of 184,000 ha (The Rice Marketing Board for the State of New South Wales 2001). The industry has a farm-gate value of approximately $350 million (AUD) and total value (export earnings, value-added) of over \$800 million (Sunrice, 2002). Including flow-on effects, it is estimated that the industry generates over \$4 billion annually benefiting regional communities and the Australian economy. Normally the crop is grown in pounded water from sowing or from the 3-leaf stage. Rice is often grown in rotation with leguminous pastures and dryland crops, which improve soil fertility and limit the need for pesticides. The number of farms growing rice is restricted because there is limited water available for irrigation. Due to the increasing demand and competing water uses, state and federal governments have established policies to control the allocation of water to all users and to meet necessary environmental requirements.

The ability of the soil to pond water without excessive accessions to the groundwater or environmental effects to other lands are factors considered in approving areas of a farm suitable for rice production. The drainage water from paddy fields is recycled to maximize the utility of irrigation water and minimize off-farm effects of the irrigation system. Soil salinisation due to rising water tables in the rice-growing areas is a major concern for the rice based farming systems in the region. Other concerns include the volume and chemical composition of drainage waters, breeding of mosquitoes and bird control. To reduce water requirements of the rice crop, researchers are trying to shorten the growth duration and increase cold tolerance of the crop to minimize pounding duration.

Pakistan

The cultivated area in Pakistan is about 20 Mha of which more than 16 Mha are irrigated. About 11 Mha of the irrigated area (i.e. 73 % of the total) is situated in Punjab Province which is in the rice-wheat agro-ecological zone of Pakistan.

The Indus Basin Irrigation System in Pakistan is the largest integrated irrigation system in the world. This irrigation system diverts approximately 123 BCM of annual river flow and spreads it over 13.5 Mha of cultivable land, of which nearly 9 Mha can be irrigated throughout the year. This controlled distribution is accomplished by means of 17 barrages and canal diversion works, 42 major canals, 6,000 km of minor canals, 600 km of link canals, and 78,000 watercourses. The total capacity is nearly 7,000 m^3/sec.

This flow is supplemented with over 150,000 tube wells, which pump 24.5 BCM/year from groundwater. Rice cultivation in Pakistan is concentrated in the central Punjab and north-western districts of Sindh, where both surface and groundwater irrigation systems are well developed. Basmati rice is the principal cash crop in the Kharif (summer) season and wheat in the Rabi (winter) season. Rice occupies about 25 % of the cultivated area in the summer monsoon season and 10 % of the total cropped area. Wheat, being the staple food, occupies 75 % of the cultivated areas in the winter season and about 38% of the total cropped area. Pakistan is among the four major rice-exporting countries, but produces only 5 million tons compared to Bangladesh (34); Myanmar (21); India (132); Japan (11); Philippines (13); Thailand (26); Vietnam (32); China (182) and the world (581 Mt).

In Pakistan, rice transplanting coincides with the onset of monsoon rains, which meet the major portion of the rice water requirement. Pakistan has a huge potential to increase rice growing on a large scale due to its relatively level terrain, heavy soils with good water holding capacity, sunny days, appropriate climatic conditions and abundant supply of farm labour. Unfortunately, inadequate supply of irrigation water at critical times of growth, lack of drainage, saline and sodic soils, low quality seeds, antiquated farm implements, imbalances in farm inputs, unsatisfactory agriculture and irrigation practices are major constraints limiting rice area growth and generally crop production in Pakistan. Since the introduction of canal irrigation, waterlogging and soil salinity have become the major problems impeding agricultural growth and development.

China

Asia produces 90 % of the world's rice in a climatic zone which can have an annual precipitation of more than 1,500 mm. China is the major rice producing country having about 30 Mha under rice paddy cultivation with a total rice yield of 190 Mt, which is 32% of world rice production. China has a long history of irrigation. Two thousand years ago, the world-famous Dujiangyan irrigation district in Sichuan Province was built and remains in use after many periods of rehabilitation and modification.

With a current irrigated area of 670,000 ha it is a highly developed economic centre of great importance to China's food production. In 1949, the irrigated area of the whole country was 16 Mha accounting for only 16% of the country's farmland, and the per-capita food consumption was 209 kg. Development in the last half century has resulted in the irrigated area growing to 53 Mha, accounting for 40% of the farmland, and the per-capita food consumption was 400 kg by the end of 1998. The total water use for irrigation gradually increased from approximately 100 BCM in 1949 to 358 BCM in 1980, after which it stabilised. Irrigation water use has been stable since 1980. Industrial and municipal water use has increased rapidly and reduced the proportion used in irrigation to 92, 80, and 65% in 1949, 1980 and 1997, respectively. The efficiency of irrigation water use and the production efficiency have progressively risen over the past several decades, especially since 1980. The average water use for irrigated agricultural was 875 mm in 1980 decreasing to 780 mm in 1997. The average specific food yield during the same period increased from 0.6 kg/m^3 to 1 kg/m^3.

Rice contributes over 39% of the total food grain production in China from 31 Mha (28% of 113 Mha which is the total agriculture area). In 1999, the total rice yield reached 200 Mt, which accounts for 39% of the total national grain production of the country (Editorial Committee of the Year Book of Chinese Agriculture, 2001). Before the 1970s, the traditional irrigation regime for rice was "continuous deep flooding irrigation". Under this regime, a low yield of rice was obtained with a large amount of water. As the industrial, urban and rural domestic water consumptions have increased continuously, there are less water resources available for irrigation year by year. These pressures on the water supply are similar to those experienced around the world and have raised the importance of water worldwide. Recognising the need to improve both water and land productivity China has introduced a Water Efficient Regimes programme (WEI) for their rice industry. Since the introduction of the programme in the 1980s many regions have adapted one of the three WEIs:

- S.W.D. Combining shallow water depth with wetting and drying
- A.W.D. Alternate wetting and drying
- S.D.C. Semi dry cultivation.

By 1997, 5.7 Mha had been converted to one or other of these systems. A reduced environmental impact has been claimed for each method.

Climatic, Water Balance and Hydrological Comparison of Three Subsystems

The key water balance components are compared for three irrigation areas, viz., Murrumbidgee Irrigation Area (MIA), Rechna Doab (RD) and the Liuyuankou Irrigation System (LIS). The MIA has an arid climate with low rainfall whereas Rechna Doab and LIS both receive considerable rainfall. In all cases rainfall is clearly below potential evapotranspiration.

Australia – The Murrumbidgee Irrigation Area

The Murrumbidgee Irrigation Area (MIA) is situated in the central New South Wales region of south-east Australia. Irrigation suitability studies were undertaken along the Murrumbidgee River in the 1890s, with development taking place between 1906 and 1913. By 1914, there were 677 farms in the MIA. Water was supplied by the first major reservoir built for irrigation – Burrinjuck Dam, which was completed in 1924. Rice growing started in the MIA in 1924, although rapid development of rice areas occurred in the 1970s and 1980s. The total area for the MIA is 156,605 ha and the main agricultural products are rice, grapes and citrus. Rice is the most dominant water user with more than 32,000 ha (14 % of the total landscape) in 2000. Irrigation demand for crop production is mainly met by water drawn from the channels and less so by groundwater pumping

These inputs and outputs have been aggregated from a spatial model (750m grid) of the MIA to provide a lumped water balance for discussion in this paper. From the point of view of overall sustainability, it is necessary to consider total flow to and from the aquifer system in relation to total lateral outflow potential of the system. If there is continuous accumulation of flows and salts then the area will eventually become waterlogged and salinised. This has already happened in many of the horticultural areas, which have been under an intensive irrigation regime for 60 years and therefore needed artificial drainage.

The total (lateral+ pumping) outflow of the aquifer, within the considered spatial boundaries of the system, is less than the total (vertical recharge and lateral) inflow to the aquifer. If these trends continue the groundwater pressure will rise and overall rate of soil salinisation will increase. This will result in substantial yield decline over time, environmental degradation within and outside the area, depreciation of natural capital and therefore a corresponding reduction in water use benefits from the system. For long term sustainability of this system there is a need to export salts from the area by establishing surface and subsurface drainage.

China – The Liuyuankou Irrigation System

The Liuyuankou Irrigation System (LIS) is located in Kaifeng County in the Chinese province of Henan and has been operational since 1967. The major crops are maize, rice and cotton. Eight branch channels were constructed between 1984 and 1988. Irrigation for crop production is met by water drawn from the channels as well as by groundwater pumping. During recent years irrigation conditions have become more efficient due to the improvement and maintenance of the hydraulic structures. In spite of this improved efficiency and the presence of a drainage system, the groundwater table has risen alarmingly. In the northern part of the LIS the groundwater tables are very shallow, within 1 m of the land surface. The lateral outflow of the aquifer is very small compared to the total inflow of water (within the considered spatial boundaries of the system), a significant amount of the irrigation water leaves LIS through fallow evaporation and crop transpiration.

A significant component of the overall water balance is the large lateral seepage from the Yellow River, which equates to an overall water supply of a similar order to the surface diversion. This is attributed to greater hydraulic conductivity of the aquifers and the height of the Yellow River above the surrounding plains. The lateral outflow from the aquifer, within the considered spatial boundaries of the system, is very small compared with the total vertical recharge and lateral inflow to the aquifer.

The groundwater aquifer is already full and there is risk of soil salinisation if hydraulic loading due to rice is reduced as it is mainly responsible for pushing salts down through the aquifer system. Since the overall outflow is very small this area is a net salt sink and is recycling these salts through the system by groundwater pumping. This feature of the system could cause substantial yield decline in the future. There is need to change groundwater pumping to shallow watertable areas and introduce more surface water supplies in the present groundwater-dependent areas.

Pakistan – The Rechna Doab

The Rechna Doab ("land between two rivers") is the interfluvial sedimentary basin of the Chenab and Ravi rivers in Pakistan. It is one of the oldest, agriculturally-richest and most intensively-populated irrigated areas of Punjab Province. Irrigation water is pumped from the aquifer and drawn from the rivers. The gross area of Rechna Doab is 2.97 Mha, with a longitudinal extent of 403 km and a maximum width of 113 km. The area falls in the rice-wheat and sugarcane-wheat agro-climatic zones of the Province, with rice, cotton and forage crops dominating in summer, wheat and forage in winter. In some parts, sugar cane is also cultivated as an annual crop.

In the upper part of Rechna Doab a significant amount of surface water is used for irrigation. Relatively low volumes of surface water are available in the lower part of Rechna Doab, therefore crop demand is met to a larger extent by groundwater pumping. The declining groundwater levels in the lower part of Doab are reducing profitability as costs for pumping groundwater increase. In contrast to the upper Rechna Doab, there is no evaporation from fallowed soil in the lower Doab. However, salinity is a bigger problem there because the groundwater used for irrigation contains increasing concentration of salt in water leached water below the root zone of crops. Larger scale drawdown of the watertable in the lower part of Doab is also promoting lateral flow of saline groundwater from its central part.

The challenges faced by present day irrigated agriculture are not much different to those faced by ancient agricultural systems such as those of Mesopotamia. A close examination of the water balance components of three irrigation systems in Australia, China and Pakistan show that surface water efficiency is highest for the Murrumbidgee Irrigation Area (over 77 %) whereas surface water efficiency of Rechna Doab in Pakistan and LIS in China are less than 50 %. All these systems are dependent on direct or indirect use of groundwater.

The direct shallow groundwater uptake by crops is very high in the LIS and MIA. If the direct groundwater use by crops continues in both the MIA and LIS, it will accelerate the rate of salinisation of soils. In the case of Rechna Doab more than 50% of crop water requirements are met from groundwater pumping which may result in salinity and sodicity unless adequate leaching of the root zone is not maintained. A key question for systems such as Rechna Doab and LIS is whether it is more cost effective to reduce seepage from channels or to pump groundwater.

There is a need to quantify regional water quality trends, downstream environmental impacts and the trade-off between yield reduction and direct regional groundwater use by crops in these systems. We can maintain the productive function of any of these areas by providing adequate drainage and salt export facilities, which have high energy and capital requirements. The most cost-effective option may be to increase water use efficiency and reduce negative impacts on the environment thereby reducing the associated costs of maintaining our natural capital. There is a need to radically rethink sustainability of food production, rational pricing and sharing of water and commodities to justify investment that will maintain and enhance ecosystem function within irrigated catchments. Under present operational conditions none of the three systems discussed in this paper is sustainable.

Types of Irrigation

Various types of irrigation techniques differ in how the water obtained from the source is distributed within the field. In general, the goal is to supply the entire field uniformly with water, so that each plant has the amount of water it needs, neither too much nor too little.

Surface Irrigation

Surface irrigation is defined as the group of application techniques where water is applied and distributed over the soil surface by gravity. It is by far the most common form of irrigation throughout the world and has been practiced in many areas virtually unchanged for thousands of years.

Surface irrigation is often referred to as flood irrigation, implying that the water distribution is uncontrolled and therefore, inherently inefficient. In reality, some of the irrigation practices grouped under this name involve a significant degree of management (for example surge irrigation). Surface irrigation comes in three major types; level basin, furrow and border strip.

The Process

The process of surface irrigation can be described using four phases. As water is applied to the top end of the field it will flow or advance over the field length. The advance phase refers to that length of time as water is applied to the top end of the field and flows or advances over the field length. After the water reaches the end of the field it will either run-off or start to pond. The period of time between the end of the advance phase and the shut-off of the inflow is termed the wetting, pending or storage phase. As the inflow ceases the water will continue to runoff and infiltrate until the entire field is drained. The depletion phase is that short period of time after cut-off when the length of the field is still submerged. The recession phase describes the time period while the water front is retreating towards the downstream end of the field. The depth of water applied to any point in the field is a function of the opportunity time, the length of time for which water is present on the soil surface.

Basin Irrigation

Level basin irrigation has historically been used in small areas having level surfaces that are surrounded by earth banks. The water is applied rapidly to the entire basin and is allowed to infiltrate. Basins may be linked sequentially so that drainage from one basin is diverted into the next once the desired soil water deficit is satisfied. A "closed" type basin is one where no water is drained from the basin. Basin irrigation is favoured in soils with relatively low infiltration rates (Walker and Skogerboe 1987). Fields are typically set up to follow the natural contours of the land but the introduction of laser levelling and land grading has permitted the construction of large rectangular basins that are more appropriate for mechanised broadacre cropping. Basin irrigation is commonly used in the production of crops such as rice and wheat.

Furrow Irrigation

Furrow irrigation is conducted by creating small parallel channels along the field length in the direction of predominant slope. Water is applied to the top end of each furrow and flows down the field under the influence of gravity. Water may be supplied using gated pipe, siphon and head ditch or bankless systems. The speed of water movement is determined by many factors such as slope, surface roughness and furrow shape but most importantly by the inflow rate and soil infiltration rate.

The spacing between adjacent furrows is governed by the crop species, common spacings typically range from 0.75 to 2 metres. The crop is planted on the ridge between furrows which may contain a single row of plants or several rows in the case of a bed type system. Furrows may range anywhere from less than 100 m to 2000 m long depending on the soil type, location and crop type. Shorter furrows are commonly associated with higher uniformity of application but result in increasing potential for runoff losses. Furrow irrigation is particularly suited to broad-acre row crops such as cotton, maize and sugar cane. It is also practiced in various horticultural industries such as citrus, stone fruit and tomatoes.

The water can take a considerable period of time to reach the other end, meaning water has been infiltrating for a longer period of time at the top end of the field. This results in poor uniformity with high application at the top end with lower application at the bottom end. In most cases the performance of furrow irrigation can be improved through increasing the speed at which water moves along the field (the advance rate). This can be achieved through increasing flow rates or through the practice of surge irrigation. Increasing the advance rate not only improves the uniformity but also reduces the total volume of water required to complete the irrigation.

Surge Irrigation

Surge Irrigation is a variant of furrow irrigation where the water supply is pulsed on and off in planned time periods (e.g. on for ½ hour off for ½ hour). The wetting and drying cycles reduce infiltration rates resulting in faster advance rates and higher uniformities than continuous flow. The reduction in infiltration is a result of surface consolidation, filling of cracks and micro pores and the disintegration of soil particles during rapid wetting and consequent surface sealing during each drying phase. The effectiveness of surge irrigation is soil type dependent, for example many clay soils experience a rapid sealing behaviour under continuous flow therefore surge offers little benefit.

Bay/Border Strip Irrigation

Border strip or bay irrigation could be considered as a hybrid of level basin and furrow irrigation. The borders of the irrigated strip are longer and the strips are narrower than for basin irrigation and are orientated to align lengthwise with the slope of the field. The water is applied to the top end of the bay, which is usually constructed to facilitate free-flowing conditions at the downstream end. One common use of this technique includes the irrigation of pasture for dairy production.

Drainage after Harvest or in Rainy Season

Drainage of flooded banks or drainage of extremely wet soil during the rainy season may be done by ditches. Drainage by ditches may be done with crops that require the soil to be wet but not completely saturated (and sometimes, especially not at certain times of year). An example is blueberries. In the rainy season/winter, they require drier soil.

Issues Associated with Surface Irrigation

While surface irrigation can be practiced effectively using the right management under the right conditions, it is often associated with a number of issues undermining productivity and environmental sustainability

- Waterlogging-Can cause the plant to shut down delaying further growth until sufficient water drains from the rootzone.

Waterlogging may be counteracted by drainage and watertable control.

- Deep drainage-Over irrigation may cause water to move below the root zone resulting in rising water tables. In regions with naturally occurring saline soil layers (for example salinity in south eastern Australia) or saline aqifers, these rising water tables may bring salt up into the root zone leading to problems of irrigation salinity.

- Salinization-Depending on water quality irrigation water may add significant volumes of salt to the soil profile. While this is a lesser issue for surface irrigation compared to other irrigation methods (due to the comparatively high leaching fraction), lack of subsurface drainage may restrict the leaching of salts from the soil. This can be remedied by drainage and soil salinity control.

Localized Irrigation

Localized irrigation is a system where water is distributed under low pressure through a piped network, in a pre-determined pattern, and applied as a small discharge to each plant or adjacent to it. Drip irrigation, spray or micro-sprinkler irrigation and bubbler irrigation belong to this category of irrigation methods.

Drip Irrigation

Drip irrigation, also known as trickle irrigation or micro irrigation, is an irrigation method which saves water and fertilizer by allowing water to drip slowly to the roots of plants, either onto the soil surface or directly onto the root zone, through a network of valves, pipes, tubing, and emitters.

History

Drip irrigation has been used since ancient times when buried clay pots were filled with water, which would gradually seep into the grass. Modern drip irrigation began its development in Afghanistan in 1866 when researchers began experimenting with irrigation using clay pipe to create combination irrigation and drainage systems. In 1913, E.B. House at Colorado State University succeeded in applying water to the root zone of plants without raising the water table. Perforated pipe was introduced in Germany in the 1920s and in 1934, O.E. Nobey experimented with irrigating through porous canvas hose at Michigan State University.

With the advent of modern plastics during and after World War II, major improvements in drip irrigation became possible. Plastic microtubing and various types of emitters began to be used in the greenhouses of Europe and the United States.

The modern technology of drip irrigation was invented in Israel by Simcha Blass and his son Yeshayahu. Instead of releasing water through tiny holes, blocked easily by tiny particles, water was released through larger and longer passageways by using velocity to slow water inside a plastic emitter. The first experimental system of this type was established in 1959 when Blass partnered with Kibbutz Hatzerim to create an irrigation company called Netafim. Together they developed and patented the first practical surface drip irrigation emitter. This method was very successful and subsequently spread to Australia, North America, and South America by the late 1960s.

In the United States, in the early 1960s, the first drip tape, called *Dew Hose*, was developed by Richard Chapin of Chapin Watermatics (first system established during 1964). In Pakistan it has been promoted by the Pakistan Atomic Energy Commission, the Agriculture Development Bank as well as successive governments. Beginning in 1989, Jain irrigation helped pioneer effective water-management through drip irrigation in India. Jain irrigation also introduced some drip irrigation marketing approaches to Indian agriculture such as 'Integrated System Approach', One-Stop-Shop for Farmers, 'Infrastructure Status to Drip Irrigation & Farm as Industry.' The latest developments in the field involve even further reduction in drip rates being delivered and less tendency to clog. One of the prestigious names in the field of drip irrigation in India is Kisan Irrigations Limited. Although among the top names in the field of Pipes & Fittings in India, they were a late entrant in this field. However, they have made important strides with their innovative & futuristic products like Hydrozig & Flat ranges in driplines as well as semi-portable, portable & flexible Mini and Micro Sprinkler Irrigation Systems for farmers in India. Their systems have been widely accepted by farmers due to the lowest clogging rate in operational conditions mainly due to the efficient Filtration Unit used which are indigenously manufactured at Kisan.

Modern drip irrigation has arguably become the world's most valued innovation in agriculture since the invention of the impact sprinkler in the 1930s, which offered the first practical alternative to surface irrigation. Drip irrigation may also use devices called micro-spray heads, which spray water in a small area, instead of dripping emitters. These are generally used on tree and vine crops with wider root zones. Subsurface drip irrigation (SDI) uses permanently or temporarily buried dripperline or drip tape located at or below the plant roots. It is becoming popular for row crop irrigation, especially in areas where water supplies are limited or recycled water is used for irrigation. Careful study of all the relevant factors like land topography, soil, water, crop and agro-climatic conditions are needed to determine the most suitable drip irrigation system and components to be used in a specific installation.

Components and Operation

Components (listed in order from water source):

- Pump or pressurized water source
- Water Filter(s)-Filtration Systems: Sand Separator like Hydro-Cyclone, Screen filters, Media Filters
- Fertigation Systems (Venturi injector) and Chemigation Equipment (optional)
- Backwash Controller (Backflow Preventer)
- Pressure Control Valve (Pressure Regulator)
- Main Line (larger diameter Pipe and Pipe Fittings)
- Hand-operated, electronic, or hydraulic Control Valves and Safety Valves
- Smaller diameter polytube (often referred to as "laterals")
- Poly fittings and Accessories (to make connections)
- Emitting Devices at plants (ex. Emitter or Drippers, micro spray heads, on-line drippers, trickle rings)
- Note that in Drip irrigation systems Pump and valves may be manually or automatically operated by a controller.

Most large drip irrigation systems employ some type of filter to prevent clogging of the small emitter flow path by small waterborne particles. New technologies are now being offered that minimize clogging. Some residential systems are installed without additional filters since potable water is already filtered at the water treatment plant. Virtually all drip irrigation equipment manufacturers recommend that filters be employed and generally will not honour warranties unless this is done. Last line filters just before the final delivery pipe are strongly recommended in addition to any other filtration system due to fine particle settlement and accidental insertion of particles in the intermediate lines.

Drip and subsurface drip irrigation is used almost exclusively when using recycled municipal waste water. Regulations typically do not permit spraying water through the air that has not been fully treated to potable water standards. Because of the way the water is applied in a drip system, traditional surface applications of timed-release fertilizer are sometimes ineffective, so drip systems often mix liquid fertilizer with the irrigation water. This is called fertigation; fertigation and chemigation (application of pesticides and other chemicals to periodically clean out the system, such as chlorine or sulfuric acid) use chemical injectors such as diaphragm pumps, piston pumps, or venturi pumps.

The chemicals may be added constantly whenever the system is irrigating or at intervals. Fertilizer savings of up to 95% are being reported from recent university field tests using drip fertigation and slow water delivery as compared to timed-release and irrigation by micro spray heads. If properly designed, installed, and managed, drip irrigation may help achieve water conservation by reducing evaporation and deep drainage when compared to other types of irrigation such as flood or overhead sprinklers since water can be more precisely applied to the plant roots. In addition, drip can eliminate many diseases that are spread through water contact with the foliage. Finally, in regions where water supplies are severely limited, there may be no actual water savings, but rather simply an increase in production while using the same amount of water as before. In very arid regions or on sandy soils, the preferred method is to apply the irrigation water as slowly as possible. Pulsed irrigation is sometimes used to decrease the amount of water delivered to the plant at any one time, thus reducing runoff or deep percolation. Pulsed systems are typically expensive and require extensive maintenance. Therefore, the latest efforts by emitter manufacturers are focused toward developing new technologies that deliver irrigation water at ultra-low flow rates, i.e. less than 1.0 liter per hour. Slow and even delivery further improves water use efficiency without incurring the expense and complexity of pulsed delivery equipment.

Drip Irrigation is Used by Farms, Commercial Greenhouses and Residential Gardeners

Drip irrigation is adopted extensively in areas of acute water scarcity and especially for crops such as coconuts, containerized landscape trees, grapes, bananas, ber, brinjal, citrus, strawberries, sugarcane, cotton, maize, and tomatoes.

Garden

Garden drip irrigation kits are increasingly popular for the homeowner and consist of a timer, hose and emitter. Hoses that are 4 mm in diameter are used to irrigate flower pots.

Advantage/Disadvantages

The advantages of drip irrigation are:

- Minimized fertilizer/nutrient loss due to localized application and reduced leaching.
- High water application efficiency.
- Levelling of the field not necessary.
- Ability to irrigate irregular shaped fields.
- Allows safe use of recycled water.
- Moisture within the root zone can be maintained at field capacity.
- Soil type plays less important role in frequency of irrigation.
- Minimized soil erosion.
- Highly uniform distribution of water i.e., controlled by output of each nozzle.
- Lower labour cost.
- Variation in supply can be regulated by regulating the valves and drippers.
- Fertigation can easily be included with minimal waste of fertilizers.
- Foliage remains dry thus reducing the risk of disease.
- Usually operated at lower pressure than other types of pressurised irrigation, reducing energy costs.

The disadvantages of drip irrigation are:

- Expense. Initial cost can be more than overhead systems.
- Waste. The sun can affect the tubes used for drip irrigation, shortening their usable life. Longevity is variable.
- Clogging. If the water is not properly filtered and the equipment not properly maintained, it can result in clogging.
- Drip irrigation might be unsatisfactory if herbicides or top dressed fertilizers need sprinkler irrigation for activation.

- Drip tape causes extra cleanup costs after harvest. You'll need to plan for drip tape winding, disposal, recycling or reuse.
- Waste of water, time & harvest, if not installed properly. These systems requires careful study of all the relevant factors like land topography, soil, water, crop and agro-climatic conditions, and suitability of drip irrigation system and its components.
- Germination Problems. In lighter soils subsurface drip may be unable to wet the soil surface for germination. Requires careful consideration of the installation depth.
- Salinity. Most drip systems are designed for high efficiency, meaning little or no leaching fraction. Without sufficient leaching, salts applied with the irrigation water may build up in the root zone, usually at the edge of the wetting pattern.

Dripperline

A dripperline is a type of drip irrigation tubing with emitters pre-installed at the factory.

Emitter

An emitter is also called a dripper and is used to transfer water from a pipe or tube to the area that is to be irrigated. Typical emitter flow rates are from 0.16 to 4.0 US gallons per hour (0.6 to 16 L/h). In many emitters, flow will vary with pressure, while some emitters are *pressure compensating.* These emitters employ silicone diaphragms or other means to allow them to maintain a near-constant flow over a range of pressures, for example from 10 to 50 psi (70 to 350 kPa).

Sprinkler Irrigation

In sprinkler or overhead irrigation, water is piped to one or more central locations within the field and distributed by overhead high-pressure sprinklers or guns. A system utilizing sprinklers, sprays, or guns mounted overhead on permanently installed risers is often referred to as a *solid-set* irrigation system. Higher pressure sprinklers that rotate are called *rotors* and are driven by a ball drive, gear drive, or impact mechanism.

Rotors can be designed to rotate in a full or partial circle. Guns are similar to rotors, except that they generally operate at very high pressures of 40 to 130 lbf/in^2 (275 to 900 kPa) and flows of 50 to 1200 US gal/min (3 to 76 L/s), usually with nozzle diameters in the range of 0.5 to 1.9 inches (10 to 50 mm). Guns are used not only for irrigation, but also for industrial applications such as dust suppression and logging.

Sprinklers may also be mounted on moving platforms connected to the water source by a hose. Automatically moving wheeled systems known as *travelling sprinklers* may irrigate areas such as small farms, sports fields, parks, pastures, and cemeteries unattended. Most of these utilize a length of polythene tubing wound on a steel drum. As the tubing is wound on the drum powered by the irrigation water or a small gas engine, the sprinkler is pulled across the field. When the sprinkler arrives back at the reel the system shuts off. This type of system is known to most people as a "waterreel" travelling irrigation sprinkler and they are used extensively for dust suppression, irrigation, and land application of waste water. Other travellers use a flat rubber hose that is dragged along behind while the sprinkler platform is pulled by a cable. These cable-type travellers are definitely old technology and their use is limited in today's modern irrigation projects.

Centre Pivot Irrigation

Centre-pivot irrigation (sometimes called central pivot irrigation), also called circle irrigation, is a method of crop irrigation in which equipment rotates around a pivot. A circular area centred on the pivot is irrigated, often creating a circular pattern in crops when viewed from above.

How it Works

Central pivot irrigation is a form of overhead (sprinkler) irrigation consisting of several segments of pipe (usually galvanized steel or aluminium) joined together and supported by trusses, mounted on wheeled towers with sprinklers positioned along its length. The machine moves in a circular pattern and is fed with water from the pivot point at the centre of the circle. The outside set of wheels sets the master pace for the rotation (typically once every three days). The inner sets of wheels are mounted at hubs between two segments and use angle sensors to detect when the bend at the joint exceeds a certain threshold, and thus, the wheels should be rotated to keep the segments aligned. Centre pivots are typically less than 500m in length (circle radius) with the most common size being the standard 1/4 mile machine (400 m). To achieve uniform application, centre pivots require a continuously variable emitter flow rate across the radius of the machine. Nozzle sizes are smallest at the inner spans to achieve low flow rates and increase with distance from the pivot point.

Most centre pivot systems now have drops hanging from a u-shaped pipe called a *gooseneck* attached at the top of the pipe with sprinkler heads that are positioned a few feet (at most) above the crop, thus limiting evaporative losses and wind drift. There are many different nozzle configurations available including static plate, moving plate and part circle. Pressure regulators are typically installed upstream of each nozzle to ensure each is operating at the correct design pressure. Drops can also be used with drag hoses or bubblers that deposit the water directly on the ground between crops. This type of system is known as LEPA (Low Energy Precision Application) and is often associated with the construction of small dams along the furrow length (termed furrow diking/dyking). Crops may be planted in straight rows or are sometimes planted in circles to conform to the travel of the centre pivot.

Originally, most centre pivots were water-powered. These were replaced by hydraulic systems and electric motor-driven systems. Most systems today are driven by an electric motor mounted at each tower.

For centre pivot to be used, the terrain needs to be reasonably flat; but one major advantage of centre pivots over alternative systems is the ability to function in undulating country. This advantage has resulted in increased irrigated acreage and water use in some areas. The system is in use, for example, in parts of the United States, Australia, New Zealand, Brazil and also in desert areas such as the Sahara and the Middle East.

Centre Pivot Manufacturers

There are four major centre pivot system manufacturers in the United States: Valmont Industries and their "Valley" products, Lindsay Corporation and their "Zimmatic" brand, Reinke Irrigation with their "Electrogator" machines, and T-L Irrigation who makes a hydrostatically powered system. Valley, Lindsay, and Reinke all manufacture systems powered by 480 volt electricity. T-L's variable-displacement hydraulic pump which is typically driven by a 15 horse power motor on standard quarter section irrigators.

Water application typically consist of brass impacts, drip tubes, rotating nozzles, and stationary sprays. These sprinklers are manufactured by Nelson Irrigation and Senniger Irrigation. Varying applications, soils, and crops require different volumes of water and application rates. Pivots often have a large bore impact sprinkler (called "big guns") located on the vary end of the machine to aid in irrigating most number of acres possible. While these "end guns" may dramatically increase the irrigated area they suffer from poor uniformity and may have negative impacts on the entire pivot if not designed properly.

The largest maker of mechanized irrigation market is Valmont Industries. Valmont claims a market share between 55-65% of all new centre pivots sold in the United States. Reinke is a privately held company which limits the ability for market researches to determine the exact number of centre pivot sold. Reinke and Zimmatic compete to share between 30-40% of the irrigation market. Valley, Zimmatic, and Reinke manufacture modern irrigation equipment and consume about 95% and support networks of professional dealers. T-L Irrigation, also privately held, manufactures a hydraulically driven irrigator and typically sells through part-time and farmer dealers.

Linear/Lateral Move Irrigation Machines

The above mentioned equipment can also be configured to move in a straight line where it is termed a *linear move* or *lateral move* irrigation system. In this case the water is supplied by an irrigation channel running the length of the field and positioned either at one side or midway across the field width. The motor and pump equipment is mounted on a cart adjacent to the supply channel that travels with the machine. Farmers may opt for linear moves to conform to existing rectangular field designs such as those converting from furrow irrigation. Lateral moves are far less common, rely on more complex guidance systems and require additional management than compared to centre pivot systems. Lateral moves are common in Australia and typically range between 500-1000m in length.

Lateral Move (Side Roll, Wheel Line) Irrigation

A series of pipes, each with a wheel of about 1.5 m diameter permanently affixed to its midpoint and sprinklers along its length, are coupled together at one edge of a field. Water is supplied at one end using a large hose. After sufficient water has been applied, the hose is removed and the remaining assembly rotated either by hand or with a purpose-built mechanism, so that the sprinklers move 10 m across the field. The hose is reconnected. The process is repeated until the opposite edge of the field is reached. This system is less expensive to install than a centre pivot, but much more labour intensive to operate, and it is limited in the amount of water it can carry. Most systems utilize 4 or 5-inch (130 mm) diameter aluminum pipe. One feature of a lateral move system is that it consists of sections that can be easily disconnected. They are most often used for small or oddly-shaped fields, such as those found in hilly or mountainous regions, or in regions where labour is inexpensive.

Sub-irrigation

Subirrigation also sometimes called *seepage irrigation* has been used for many years in field crops in areas with high water tables. It is a method of artificially raising the water table to allow the soil to be moistened from below the plants' root zone. Often those systems are located on permanent grasslands in lowlands or river valleys and combined with drainage infrastructure. A system of pumping stations, canals, weirs and gates allows it to increase or decrease the water level in a network of ditches and thereby control the water table.

Sub-irrigation is also used in commercial greenhouse production, usually for potted plants. Water is delivered from below, absorbed upwards, and the excess collected for recycling. Typically, a solution of water and nutrients floods a container or flows through a trough for a short period of time, 10–20 minutes, and is then pumped back into a holding tank for reuse. Sub-irrigation in greenhouses requires fairly sophisticated, expensive equipment and management. Advantages are water and nutrient conservation, and labour-saving through lowered system maintenance and automation. It is similar in principle and action to subsurface drip irrigation.

Manual Irrigation Using Buckets or Watering Cans

These systems have low requirements for infrastructure and technical equipment but need high labour inputs. Irrigation using watering cans is to be found for example in peri-urban agriculture around large cities in some African countries.

Automatic, Non-electric Irrigation Using Buckets and Ropes

Besides the common manual watering by bucket, an automated, natural version of this also exist. Using plain polyester ropes combined with a prepared ground mixture can be used to water plants from a vessel filled with water.

The ground mixture would need to be made depending on the plant itself, yet would mostly consist of black potting soil, vermiculite and perlite. This system would (with certain crops) allow to save expenses as it does not consume any electricity and only little water (unlike sprinklers, water timers,...). However, it may only be used with certain crops (probably mostly larger crops that do not need a humid environment; perhaps e.g. paprikas).

Irrigation using Stones to Catch Water from Humid Air

In countries where at night, humid air sweeps the countryside, stones are used to catch water from the humid air by condensation. This is for example practiced in the vineyards at Lanzarote.

Dry Terraces for Irrigation and Water Distribution

In subtropical countries as Mali and Senegal, a special type of terracing (without flood irrigation or intent to flatten farming ground) is used. Here, a 'stairs' is made through the use of ground level differences which helps to decrease water evaporation and also distributes the water to all patches (sort of irrigation).

Sources of Irrigation Water

Sources of irrigation water can be groundwater extracted from springs or by using wells, surface water withdrawn from rivers, lakes or reservoirs or non-conventional sources like treated wastewater, desalinated water or drainage water. A special form of irrigation using surface water is spate irrigation, also called floodwater harvesting. In case of a flood (spate) water is diverted to normally dry river beds (wadi's) using a network of dams, gates and channels and spread over large areas. The moisture stored in the soil will be used thereafter to grow crops. Spate irrigation areas are in particular located in semi-arid or arid, mountainous regions. While floodwater harvesting belongs to the accepted irrigation methods, rainwater harvesting is usually not considered as a form of irrigation. Rainwater harvesting is the collection of runoff water from roofs or unused land and the concentration of this.

How an in-ground Irrigation System Works

Most commercial and residential irrigation systems are "in ground" systems, which means that everything is buried in the ground. With the pipes, sprinklers, emitters (drippers), and irrigation valves being hidden, it makes for a cleaner, more presentable landscape without garden hoses or other items having to be moved around manually. This does, however, create some drawbacks in the maintenance of a completely buried system.

Water Source and Piping

The beginning of a sprinkler system is the water source. This is usually a tap into an existing (city) water line or a pump that pulls water out of a well or a pond. The water travels through pipes from the water source through the valves to the sprinklers and emitters.

The pipes from the water source up to the irrigation valves are called "mainlines", and the lines from the valves to the emitters or sprinklers are called "lateral lines". Most piping used in irrigation systems today are HDPE and MDPE or PVC or PEX plastic pressure pipes due to their ease of installation and resistance to corrosion. After the water source, the water usually travels through a check valve. This prevents water in the irrigation lines from being pulled back into and contaminating the clean water supply. Ideally a pressure control valve is also installed to regulate water pressure and help prevent excessive pressure from harming the system.

Controllers, Zones and Valves

Most irrigation systems are divided into zones. A zone is a single irrigation valve and one or a group of drippers or sprinklers that are connected by pipes or tubes. Irrigation systems are divided into zones because there is usually not enough pressure and available flow to run sprinklers for an entire yard or sports field at once. Each zone has a solenoid valve on it that is controlled via wire by an irrigation controller.

The irrigation controller is either a mechanical (now the "dinosaur" type) or electrical device that signals a zone to turn on at a specific time and keeps it on for a specified amount of time. "Smart Controller" is a recent term used to describe a controller that is capable of adjusting the watering time by itself in response to current environmental conditions. The smart controller determines current conditions by means of historic weather data for the local area, a soil moisture sensors (water potential or water content), rain sensor, or in more sophisticated systems satellite feed weather station, or a combination of these.

Emitters & Sprinklers

When a zone comes on, the water flows through the lateral lines and ultimately ends up at the irrigation emitter (drip) or sprinkler heads. Many sprinklers have pipe thread inlets on the bottom of them which allows a fitting and the pipe to be attached to them. The sprinklers are usually installed with the top of the head flush with the ground surface. When the water is pressurized, the head will pop up out of the ground and water the desired area until the valve closes and shuts off that zone. Once there is no more water pressure in the lateral line, the sprinkler head will retract back into the ground. Emitters are generally laid on the soil surface or buried a few inches to reduce evaporation losses.

Problems in irrigation:

- Competition for surface water rights.
- Depletion of underground aquifers.
- Ground subsidence (e.g. New Orleans, Louisiana).

Underirrigation or irrigation giving only just enough water for the plant (e.g. in drip line irrigation) gives poor soil salinity control which leads to increased soil salinity with consequent build up of toxic salts on soil surface in areas with high evaporation. This requires either leaching to remove these salts and a method of drainage to carry the salts away. When using drip lines, the leaching is best done regularly at certain intervals (with only a slight excess of water), so that the salt is flushed back under the plant's roots.

- Over irrigation because of poor distribution uniformity or management wastes water, chemicals, and may lead to water pollution.
- Deep drainage (from over-irrigation) may result in rising water tables which in some instances will lead to problems of irrigation salinity.
- Irrigation with saline or high-sodium water may damage soil structure.

Reservoir

A reservoir is an artificial lake used to store water. Reservoirs are often created by building a reinforced dam, usually out of concrete, earth, rock, or a mixture across a river or stream. Once the dam is completed, the stream fills the reservoir. When a reservoir is predominantly man-made (rather than being an adaptation of a natural basin) it may be called a cistern. The term reservoir is also often used to describe underground reservoirs such as an oil or water well.

Types

Valley Dammed Reservoir

The more common dam across a valley relies on naturally formed features to form the watertight elements. Generally, engineers look for dam sites which are narrow with a broad area upstream; the valley sides can then act as natural walls and the broad area upstream makes a large reservoir for the height. The best place along the valley for building a dam has to be determined according to where the dam can best be tied into the valley walls and floor to form a watertight seal. If necessary, humans have to be re-housed or historic sites must be moved. For example, the temples of Abu Simbel were moved before the construction of the Aswan Dam (which created Lake Nasser from the Nile in Egypt). At the start of construction, the river must be diverted, often through a tunnel.

Then the foundation is prepared. Once that is done, building of the dam can start. This may take anywhere from a few months to a few years, depending on its size and complexity. After the dam is complete, the diversion is removed or plugged, and the river fills the area upstream of the dam.

Bank-side Reservoir

Where water is taken from a river of variable quality or quantity, it is common to construct bank-side reservoirs to store water pumped or siphoned from the river. Such reservoirs are usually built partly by excavation and partly by the construction of a complete encircling bound or embankment. Both the floor of the reservoir and the bound must have an impermeable lining or core, often made of puddled clay. The water stored in such reservoirs may have a residence time of several months during which time normal biological processes are able to substantially reduce many contaminants and almost eliminate any turbidity.

The use of bank-side reservoirs also allows a water abstraction to be closed down for extended period at times when the river is unacceptably polluted or when flow conditions are very low due to drought. The London water supply system is one example of the use of bank-side storage for all the water taken from the River Thames and River Lee with many large reservoirs visible along the approach to London Heathrow Airport.

Service Reservoir

Many service reservoirs are constructed as water towers, often as elevated structures on concrete pillars where the landscape is relatively flat. Other service reservoirs are entirely underground, especially in more hilly or mountainous country. In the United Kingdom Thames Water has many underground reservoirs beneath London built in the 1800s by the Victorians, most of which are lined with thick layers of brick. Honour Oak Reservoir, which was completed in 1909, is the largest of this type in Europe. The roof is supported using large brick pillars and arches and the outside surface is used as a golf course.

Operation

A raw water reservoir does not simply hold water until it is needed. It is the first part of the water treatment process. The time the water is held for before it is released is known as the *retention time*. This is a design feature that allows particles and silts to settle out, as well as time for natural biological treatment using algae, bacteria and zooplankton that naturally live within the water.

Water can be released from the reservoir, generally by gravity, to be cleaned for drinking water, generate electricity, or simply maintain the downstream flow. In the event that major rainfall occurs, water can be released via a spillway to avoid over-topping and compromising the integrity of the dam. Most modern reservoirs have a specially designed draw-off tower that can discharge water from the reservoir at different levels both to access water as the reservoir draws down but also to allow water of a specific quality to be discharged into the downstream river as compensation water.

Levels

The terminology for reservoirs varies from country to country. In the United States the normal maximum level of a reservoir lake is called *full pool*, while the minimum level it can function at is *dead pool*. The water below this point is also called the dead pool, while the water in between is called the *conservation pool*. Full pool may have different levels in summer and winter, or based on the local wet and dry seasons.

Once a reservoir reaches dead pool, it is below the level at which the dam can release it downstream. At this point, the streambed beyond the dam goes nearly or completely dry, and electricity production stops. This is also often the point at which intakes for municipal water systems begin to suck air in, and must be extended into deeper water, where stagnant water quality is much poorer. This can be done either permanently with longer pipes, or temporarily with large hoses floated on small barges, such as until a severe drought or dam repairs are over.

Hydroelectricity

A hydroelectric power station consists of large turbines at the base of a dam. Water from the reservoir behind the dam is channelled through pipes and delivered to the turbines, which in turn, spin a generator to produce electricity.

Controlling Watercourses

Reservoirs can be used in a number of ways to control how water flows through downstream waterways.

Irrigation

Water in an irrigation reservoir is released into networks of canals mainly for use in farmlands or secondary water systems. Water in an irrigation reservoir is *generally* not used for drinking water, but in some cases is.

Flood Control

Commonly known as an *"attenuation"* or *"balancing"* reservoir, these are used to prevent flooding to lower lying lands, flood control reservoirs collect water at times of unseasonally high rainfall, then release it slowly over the course of the following weeks or months. Some of these reservoirs are constructed across the river line with the onward flow controlled by an orifice plate. When river flow exceeds the capacity of the orifice plate water builds behind the dam but as soon as the flow rate reduces the water behind the dam slowly releases until the reservoir is empty again.

In some cases such reservoirs only function a few times in a decade and the land behind the reservoir may be developed as community or recreational land. A new generation of balancing dams are being developed to combat climate change. They are called "Flood Detention Reservoirs". Because these reservoirs will remain dry for long periods, there will be a question as to the stability of the clay core as it could dry out. A British company Instant Barrage Services has developed an interesting composite core fill as an alternative to clay, made from recycled materials that seems to work and has a much lower carbon footprint.

Compensation

If a standard reservoir is built on a river which is used as a source of power, a compensation reservoir may also be built to guarantee a sufficient flow of water downstream during the working hours of the water-powered industries.

Canals

Where a natural watercourse's water is not available to be diverted into a canal, a reservoir may be built to guarantee the water level in the canal; for example, where a canal climbs to cross a range of hills through locks.

Recreation

Reservoirs often provide for recreational uses. Most reservoirs are built for a civic purpose, but still allow fishing, boating, and other activities. At most reservoirs, special rules apply for the safety of the public.

Modelling Reservoir Management

There is a wide variety of software for modelling reservoirs, from the specialist Dam Safety Program Management Tools (DSPMT) to the relatively simple WAFLEX, to integrated models like the Water Evaluation And Planning system (WEAP) that place reservoir operations in the context of system-wide demands and supplies.

History

Five thousand years ago, the craters of extinct volcanoes in Arabia were used as reservoirs by farmers for their irrigation water.

Dry climate and water scarcity in India led to early development of water management techniques, including the building of a reservoir at Girnar in 3000 BC. Artificial lakes dating to the 5th century BC have been found in ancient Greece. An artificial lake in present-day Madhya Pradesh province of India, constructed in the 11th century, covered 650 square metres (7,000 sq ft).

In Sri Lanka large reservoirs have been created by ancient Sinhalese kings in order to save the water for irrigation. The famous Sri Lankan king Parakramabahu I of Sri Lanka stated " do not let a drop of water seep into the ocean without benefitting mankind ". He created the reservoir named Parakrama Samudra (sea of King Parakrama), which has astonished archeologists.

Measuring Irrigation Water

Effective irrigation water management begins with accurate water measurement. Water measurement is required to determine both total volumes of water and flow rates pumped.

Measurement of volumes will verify that the proper amount of water is applied at each irrigation and that amounts permitted by water management districts are not exceeded.

Measurement of flow rates will help to ensure that the irrigation system is operating properly. For example, low flow rates may indicate the need for pump repair or adjustment, partially closed or obstructed valves or pipelines, or clogged drip emitters. High flow rates may indicate broken pipelines, defective flush valves, too many zones operating simultaneously, or eroded sprinkler nozzles.

Units of Measurement

Volume (V)

Volume is the amount of water measured. For irrigation purposes, volume units commonly used are the gallon, acre-inch, and acre-foot. An acre-inch is the volume of water that would be required to cover an area of 1 acre to a depth of 1 inch. The relationships among these units are:

1acre-inch (ac-in)= 27,154 gallons (gal)

1acre-foot (ac-ft)=12 acre-inches (ac-in).

Depth (D)

Depth units are sometimes used to refer to the amount of water required for irrigation. Depth units (inches) are used because soil water-holding capacity is typically measured in inches (of water) per foot (of soil depth), and irrigations are scheduled after a fraction of the soil water in the plant root zone has been depleted.

For example, assume that a typical Florida fine sand soil holds 1.0 inch of water per foot of soil at field capacity, the irrigated plant root zone depth is 2 ft, and an irrigation will be scheduled when 50% of the soil water in the root zone has been depleted. Then, the total soil water storage is calculated as 2.0 inches (D = 1.0 in/ft x 2 ft soil depth = 2.0 inches), and the amount to be applied per irrigation is 1.0 inch (D = 50% x 2 inches = 1.0 inch).

Note that the amount of water to be pumped will need to be greater than the 1.0 inch to be stored in the plant root zone because some water will be lost during application. That is, application efficiencies are always less than 100 percent because of water losses due to such factors as evaporation, wind drift, and nonuniform water application. Depth units are also convenient for comparison with rainfall depths. For example, a 1-inch rainfall would supply the same volume of water as the 1-inch irrigation in the previous example. Either the 1-inch rain or irrigation would be adequate to restore the soil water content to field capacity.

Finally, plant water use rates are typically expressed in (fractions of) inches per day. For example, if the plant ET rate in the previous example is 0.25 in/day, then the allowable soil water depletion of 1.0 inch will occur in four days (1.0 inch/0.25 inches/day = 4 days), and an irrigation would be required beginning on day 5 to avoid depleting more than 50% of the available soil water in the plant root zone.

The calculation of 1.0 inch of water to be applied means that a depth of 1.0 inch is to be applied over the entire area to be irrigated. If the irrigated area is 1.0 acre, then the volume to be applied is 1.0 ac-in (V = 1.0 inch depth x 1.0 acre = 1.0 ac-in) or 27,154 gal. Likewise, if the irrigated area is 20 acres, the volume of water required is 20 ac-in or 543,080 gal (V = 20 ac-in x 27, 154 gal/ac-in = 543,080 gal). Thus, depth units are used interchangeably with volume units because it is convenient to use depth units when referring to soil water depletion, rainfall, and plant ET rates, but volume units when referring to amounts of water permitted or pumped.

Flow Rate (Q)

Flow rate is the volume of water flowing past a given point per unit of time. The units of flow rate commonly used for irrigation purposes are gallons per minute (gal/min or gpm) and acre-inches per hour (ac-in/hr). The water management districts often use million gallons per day (mgd) for permitting of water extraction rates from a water source. The relationships between these units are:

1 gal/min = 0.00221 ac-in/hr = 0.00144 mgd

1 ac-in/hr = 453 gal/min = 0.652 mgd

1 mgd = 694 gal/min = 1.53 ac-in/hr.

Since flow rate is volume per unit time, the volumes of water applied during irrigation can be calculated if the flow rate and irrigation duration (time) are known. For example, if an irrigation pump discharges 453 gpm and irrigates for 6 hours, the volume of water applied is 163,080 gal (V= 453 gal/min x 60 min/hr x 6 hr = 163,080 gal). This is equivalent to 6.0 ac-in (V = 163,080 gal/27,154 gal/ac-in = 6.0 ac-in).

Flow rates can be converted to water application rates over the irrigated area if the size of the irrigated area is known. Water application rates from sprinkler irrigation systems are typically given in in/hr, consistent with rainfall rates. If, in the previous example, a sprinkler irrigation system applies water at 453 gpm, and the irrigated area is 4 acres, then the application rate is 0.25 in/hr (453 gpm = 1.0 ac-in/hr, and 1.0 ac-in/hr/4 acres = 0.25 in/hr).

Finally, the depth of water applied can be calculated by multiplying the application rate by the duration of irrigation. Continuing the previous example, for an application rate of 0.25 in/hr and a 6 hr irrigation duration, the gross depth of water applied would be 1.5 inches (D = 0.25 in/hr x 6 hr=1.5 inches). Note that the actual depth of water stored in the plant root zone and available for plant use would be less than 1.5 inches because of water losses during application. A typical application efficiency for sprinkler irrigation systems is 75% for Florida conditions. Thus, only 1.13 inches (1.5 inches x 75% = 1.13 inches) of the 1.5 inches pumped would be expected to be stored in the root zone and available for plant use.

Velocity (v)

Velocity is the average speed at which water moves in the direction of flow. The velocity unit commonly used is feet per second (ft/sec or fps). To aid in understanding the size of this unit, it is compared with a more familiar unit:

1 ft/sec = 0.68 miles per hour (mph)

To avoid excessive pressure losses due to friction and excessive potentially damaging surge pressures, most irrigation systems are designed to avoid velocities that exceed 5 ft/sec:

5 ft/sec = 3.4 mph.

Irrigation water velocities and flow rates are sometimes mistakenly used interchangeably. Velocity and flow rate are two different (but related) concepts. Their relationship is given by the equation of continuity, a fundamental physical law:

Q = a v

where

Q= flow rate,

a = cross-sectional area of flow, and

v = velocity.

The equation of continuity states that the flow rate can be calculated from the multiple of the velocity times the cross-sectional area of flow.

In pipes flowing full, the cross-sectional area of flow is the cross-sectional area of the pipe. In open channel flow, it is the cross sectional area of the channel or stream. Thus, the equation of continuity simply states that in addition to the velocity of flow, the size of the stream or pipe also affects the flow rate.

As an example of the application of the equation of continuity, if the velocity is the same in a 1-inch and a 2-inch diameter pipe, the flow rate from the 2 inch pipe would be four times as large as the flow rate from the 1-inch diameter pipe. Note that the cross-sectional area is proportional to the diameter squared and 1 inch squared = 1 inch x 1 inch = 1 square inch, while 2 inches squared = 2 inches x 2 inches = 4 square inches. From this example, doubling the pipe diameter increases the carrying capacity of a pipe by a factor of 4.

As another example, if the previously-described 1-inch and 2-inch diameter pipes must both convey the same flow rate, then the velocity in the 1-inch pipe will be 4 times greater (inversely proportional to the ratio of the cross-sectional areas) than the velocity in the 2-inch pipe.

The amount of irrigation water pumped, applied, or used can be measured in volume or depth units. These units can be used interchangeably when it is convenient to do so. That is, the volume can be calculated from the depth applied if the area is known over which the depth was applied. The pumping rate, application rate, or plant water use rate is measured in units of volume (or depth) per unit of time. The volume can be calculated from the flow rate times the duration (time) of flow. The depth of application or use can be calculated from the application or use rate times the duration (time) of application or use. Velocity refers to the speed at which water flows (distance per unit time) as opposed to flow rate (volume per unit time). Thus, the terms flow rate and velocity cannot be used interchangeably. Both the velocity and cross-sectional area of flow must be known in order to calculate flow rates.

Spate Irrigation

Spate irrigation is a type of water management, that is unique to semi-arid environments. It is found in the Middle East, North Africa, West Asia, East Africa and parts of Latin America. Flood water from mountain catchments is diverted from river beds (wadi's) and spread over large areas. Spate systems are very risk-prone. The uncertainty comes both from the unpredictable nature of the floods and the frequent changes to the river beds from which the water is diverted. It is often the poorest segments of the rural population whose livelihood and food security depends on the spate flows. Substantial local wisdom has developed in organizing spate systems and managing both the flood water and the heavy sediment loads that go along with it.

Where does one Find Spate Irrigation Systems?

Spate irrigation occurs in areas as varied as South Asia, the Middle East, West Africa, North Africa, the Horn of Africa, Central

Asia and Latin America. Estimates for the area under flood irrigation are not easy to make, as the area under spate irrigation changes from year to year and as spate irrigation has never had the amount of attention from development agencies or tax authorities, that perennial irrigation has had.

Spate irrigation is typically found in arid and semi-arid regins, where highlands border plains. It uses seasonal floods for irrigation – but as the floods differ from year to year the area served by it fluctuates widely.

The figures for these selected countries indicates an order of magnitude. Some figures are debatable. The area under spate irrigation in Eritrea is also quoted as 90,000 ha. For Pakistan – where spate irrigation is found in all four provinces – an estimate of 3.250,000 has been mentioned by other sources. Further spate irrigation is reported from North Chile and Bolivia, Iran, Afghanistan, Mauretania and Senegal, Ethiopia, Kenya and the Northwest coast of Egypt.

Is Spate Irrigation Similar to other Flood Irrigation and Water Harvesting Systems?

As spate irrigation uses seasonal floods for irrigation, it is akin but different from two other categories of flood-based irrigation systems, i.e inundation canals (that start to flow as soon as the flood in a perennial river reaches a certain level) or flood rise or recession irrigation, where a rising perenial river over tops its banks and inundates the plains alongside the river. In flood rise or recession irrigation crops are grown on the rising or receding flow or on the residual moisture. In spate irrigation instead water is diverted from normally dry river beds (wadi's) when the river is in spate. The flood water is then diverted to the fields. This may be done by free intakes, by diversion spurs or by bunds, that are build across the river bed. The flood water – typically lasting a few hours or a few days – is channelled through a network of primary, secondary and sometimes tertiary flood channels. Command areas may range from anything between a few hectares to over 25,000 hectares.

How are Spate Irrigation Systems Managed?

Some of the larger spate irrigation rank among the largest farmer managed irrigation systems in the world. The structures are sometimes spectacular: earthen bunds, spanning the width of a river, or extensive spurs made of brushwood and stones. Spate systems are made in such a way that ideally the largest floods are kept away from the command area. Very large floods would create considerable damage to the command area. They would destroy flood diversion channels and cause rivers to shift. This is where the ingenuity of many of the traditional systems comes in. Spurs and bunds are generally made in such a way that the main diversion structures in the river break when floods are too big. Breaking of diversion structures also serves to maintain the flood water entitlements of downstream land owners.

What is the Importance of Sedimentation?

Sedimentation is another important feature of spate irrigation and spate irrigation is as much about managing water as it is about managing sedimentation. The spate waters are usually laden with sediment. Scour and siltation are part and parcel of spate irrigation. Rivers in spate lift and deposit huge quantities of sediment. As a result there is constant change in bed levels, resulting in changes in bed levels and water distribution. The impact of these processes differs between the various systems. It depends on the amount and composition of the sediment load that a river carries, which depends on the rainfall pattern and the characteristics of the catchment area; its geology, morphology and vegetation cover. Farmers are usually able to identify the origin of a flood by the type of sediment that is transported by it. The degree of siltation and scour also depends on the local topography and the type of material. In spate irrigated areas with low gradients, as are found on the plains, a river is always in danger of choking itself with its own silt deposits and finding another way. Moreover, in the fine sandy deposits of the plains, the scouring of the riverbed is a larger danger than it is in the armoured river beds of the highlands. As a result, the lowland flood irrigation systems are particularly dynamic.

Farmers, however, are not passive actors in these scour and siltation processes. They actively manipulate the scour and sedimetnation processes. They may deepen the headreach of a flood channel, in order to attract a larger flood that will further scour out the channel. If a flood river breaks its banks, farmers may close the breaches, if it deflects water away from their land or on other occasions, they will leave the breaches intact, so that these will act as escapes, creaming off the peaks of the very high floods and maintaining the flow at their own system at a manageable level. In other cases farmers will manipulate the siltation process to force the river bed to purposely silt up. The latter is in practice where the river has become uncontrollable, because its bed may has become to deep or to steep.

The remedy is to built a strong permanent bund across the river and force the river to deposit its sediment load upstream of the bund.

What is the History of Spate Irrigation?

Spate irrigation has a long history. Several sources assume that in Yemen spate irrigation started when the wet climate of the Neolithic gave way to more arid circumstances and that spate irrigation thus has been in use for five thousand years. Similarly, archeologists have discovered the remains of checkdams for spate rivers in Tauran, Iran and Balochistan, Pakistan. In Yemen spate irrigation witnessed its zenith during the Shebean period in the first millenium BC. The great Mar'ib Dam, constructed on Wadi Dhana, irrigated two oases on either banks, estimated to cover 9600 ha.

One can only speculate how the technique spread across the world. The intense development of trade after the Islamic period may have helped spread innovations from the Yemen area. The recent development of spate irrigation in Eritrea is for instance traced back to the arrival of Yemeni migrants 80-100 years ago. Yet it is likely that spate irrigation technology has sprung up independently in several areas – particularly as it is found in areas as diverse and remote as West Africa, Arabia, Central Asia and Latin America.

What is the Future of Spate Irrigation?

As a testimony of the diversity in development in the world, spate irrigation on the decline in rich areas such as Saudi Arabia, but is on the increase in low income countries such as Ethiopia and Eritrea. Generally spate irrigation is associated with low returns per labour, great variability in income between good and bad years and a high degree of social organisation to maintain the systems. Where more rewarding sources of income come up, where a period of long droughts force people to abandon their area or where the local organisation is undermined spate irrigation systems may disappear. Another important change in several areas, that are traditionally spate irrigated is the introduction of groundwater irrigation. In many spate irrigated areas groundwater resources are relatively rich due to long periods of recharge. With the availability of relatively inexpensive pumpsets groundwater has become an important source of irrigation, for instance in spate areas in Dera Ghazi Khan (Pakistan), Tunesia or Yemen. This has resulted in a neglect of the spate infrastructure and a change towards perennial cropping.

The number of public programs to support spate irrigation have been relatively limited. One reason has been the difficulty to justify investments in civil engineering works on systems, dominated with low value farming. The second reason has been that it has been hard to identify successful interventions in spate systems, because spate systems are often hydraulically and socially generally very complex.

An alternate approach to support spate systems has been the subsidization of mechanical traction. This approach has been followed with a relative high degree of success in Pakistan and Tunesia. Bulldozer programs have put a very useful resource at the hand of local spate farmers – who have remained in charge of the design and implementation. The cost effectiveness of bulldozer has been relatively high, moreover.

System Captives: Change and Stagnation in Farmer-managed Water Delivery Schedules

This paper discusses water delivery schedules in farmer managed irrigation systems, using examples from Baluchistan (Pakistan). The assumption is sometimes made that, when resource management is user controlled, changes towards more efficient use will come about automatically. The examples from Baluchistan demonstrate that such changes do not necessarily occur and correlates the lack of change to the specificity of the schedules, the different gains from the change and the importance of objectives other than improved irrigation water management.

To this day water delivery schedules continue to be a 'big unknown' in irrigation management. This is unfortunate, because their potential for increasing agricultural production from ever scarcer water is substantial. One reference (Reidinger, 1974) has described how agricultural production losses occurred as a result of the unpredictability of water supplies under a particular system of canal rotation in Hissar District in Haryana (India).

The erratic surface water deliveries were at variance with the cultivation of modern varieties and prompted farmers to develop private tubewells. Similarly, farmers had recourse to conjunctive irrigation in Leyyah District in Punjab Province (Pakistan), after a restricted canal rotation schedule was introduced in the 1970s. Yet in contrast to Hissar, in Leyyah the reduction in canal supplies and the development of tubewells had a beneficial effect, lowering groundwater tables and removing waterlogging.

Water delivery schedules deserve more attention than they receive at present. Regrettably, the discussion on irrigation management in the last few years has been overshadowed by other themes, in particular privatization, irrigation management transfer and water markets. With regard to the latter, several authors, such as Shah *et al.* (1993), argue that commercial water transactions will contribute to efficient water use by facilitating the reallocation of water to the user that is willing to offer the highest price.

This is often assumed to be the most efficient producer. To make the transactions possible, non-land-based water rights are considered essential. For permanent water transfers this may indeed be most appropriate. (This statement overlooks the informal temporary exchanges of water turns, common throughout the world of irrigation, that take place in the absence of a formalized water rights system. Similarly, informal permanent water transactions take place independently of alienable water rights and occur even in systems where water transactions are legally forbidden.) Yet water transactions are similarly facilitated or frustrated by water delivery schedules. The scope for water sales in the rice-based irrigation systems of east Asia for instance is limited, as no one can extract himself from the rigid field-to-field delivery schedule.

The relevance of water delivery schedules goes beyond whether or not they facilitate water transactions. Reliability and appropriateness of water deliveries in terms of timing, volume and flexibility are important assets. They represent social capital. In this paper, water delivery schedules in three small-scale farmer-managed systems in Baluchistan Province (Pakistan) are discussed. In each of these systems water is very scarce, the rights to it are individually owned and the systems are entirely user controlled and managed.

The water delivery schedules, unlike in large scale systems, are determined by the users themselves. With compelling simplicity, several authors such as Korten (1990) and Pretty and Guyt (1992), have argued that such local control of productive resources facilitates efficient use. The same theme surfaces in the discussion on irrigation management transfer: under farmer management, water utilization will supposedly be more efficient. Yet, is this true and to what extent? Do local institutions capture the opportunities for improvement under local control? This paper discusses institutional change in water delivery schedules in the selected farmer-managed systems. It reviews, if there is obvious scope for improvement, acknowledged by the users, whether these improvements do or do not occur. It explores what factors encourage adjustments and what factors cause, to use a term introduced by North (1990), institutional 'lock-ins'.

Farmer-managed Water Delivery Schedules

The attributes of water delivery schedules encountered in Baluchistan resemble those in small scale systems in other arid parts of the world. The private property right to the shared source of water prevails in these systems and water delivery schedules are essentially coordinating institutions between different individual water users. Individual ownership in a collective resource is symbolized by the water share, i.e., the time unit that describes the duration of a person's water turn in the total flow or in part of the flow.

In these systems three parameters describe local water delivery schedules: the duration of the irrigation cycle, the delivery pattern and the method of water distribution. The irrigation cycle is the period over which a water turn returns. Ideally, the duration of a cycle is in line with crop requirements. Irrigation cycles that are too long complicate the cultivation of crops with short irrigation intervals.

The second parameter, the delivery pattern, arranges the sequence in which each water user will receive his turn within the irrigation cycle. Ideally, the movement of water through the command area is systematic, going from head to tail or reverse rather than haphazard. This reduces losses due to dead storage and perimeter wetting. Yet where water users have scattered holdings or exchange water turns a systematic schedule may be difficult to achieve.

The final parameter that describes water delivery schedules is the method of water distribution. Water distribution directly affects discharges at field level. It divides the flow in different proportions. For this reason it is an important determinant of field efficiency as flows at field level may be too large to handle or too small to distribute across a plot. It similarly affects conveyance efficiency, since losses from small flows are proportionally larger. Another element of local water distribution is storage. Large reservoirs act as buffers. They make system discharge independent of micro-variations in inflow. Small buffers facilitate night storage and affect the timing of the supplies. They also concentrate flows, combining day and night discharges and thus help to reduce conveyance losses.

While the irrigation cycle, the delivery pattern and the distribution system are the determinants, water delivery schedules also vary in degree of specificity. In this respect there are two opposite and interchangeable principles, coordination of water deliveries by rule or by management. In the first case the irrigation cycle, flow distribution system and delivery pattern are rigidly described and are followed nearly automatically, each water user knowing his entitlement from the system. In other systems, on the other hand, an irrigation manager is in place and the discretion given to this specialist (or panel of specialists) reduces the weight and specificity of pre-set rules. These different degrees of specificity of water delivery schedules raise a number of intriguing issues.

Posner (1973) and Libecap (1989) have argued that, when the value of a resource increases, institutions become more specific and access rules will be narrowly described. This leads to a paradox: with increased specificity of rules, resource allocation becomes almost mechanistic. There will be less need for a coordinating organization. The system that comes into being is one of minimal transaction costs. However, without an actively coordinating organization, modifications will be more difficult to initiate and sub-optimal allocations are a likely consequence. This is reminiscent of a point raised by Langlois (1986) that institutions should not only be judged by their efficiency in terms of minimal transaction costs, but also by their capacity to change and be changed. A very rigid and specific schedule will not be changed easily; instead of a revision the schedules will be kept acceptable by muddling through, primarily inter-individual adjustments, such as exchanges and transfers of water turns between two users or within a subgroup of users.

In this article specificity is defined as the level of detail of resource allocation rules. Management by rules is considered the 'highly specific' extreme of a continuum that starts with active management. Alternatively, one may also argue that within the category of active management there are different poles, with the degree of specificity increasing as the decisionmaking powers are more narrowly defined.

In comparing water delivery schedules, the grounds on which to judge their performance need to be explained. There are several facets: the degree to which the water delivery schedule facilitates exchanges to high productivity users; the degree to which it optimizes the overall water utilization; and, in detail, the extent to which the water delivery schedule contributes to high field efficiency by making appropriate deliveries available; and the extent to which it minimizes conveyance and seepage losses.

Can the different facets be covered by one delivery schedule? The answer to this is dependent on the type of irrigation infrastructure in place and illustrates the importance of the artefact underneath the institution. In a system with unlined or partly lined irrigation channels, with significant conveyance losses, such as described in the three case studies below, two objectives contradict as flexibility occurs at the cost of conveyance efficiency. The frequent rearrangement of the distribution pattern that follows from commercial transactions will translate in more seepage loss as channels will be filled and emptied frequently, involving the repeated wetting of the channel perimeters.

This situation can be contrasted with a system, designed by Merriam, which consists of a network of semi-closed pipes, that distribute water that is stored in a reservoir. From a valve on their lands, farmers can take water whenever they need. Because of the pipelines, there is no conveyance loss. Similarly, flexibility is served as each farmer can take water when desired. Refreshingly, Merriam postulates that the unconstrained availability of the resource is essential for the efficient use of land *and* water. This contradicts the economist's wisdom that efficient use comes with scarcity. The weak spot in Merriam's system, however, is field irrigation efficiency, as the unrestrained availability of water does not encourage efficient field water use and, apart from self-restraint, there is no incentive not to overirrigate. Even though it is difficult to combine optimum flexibility, optimum conveyance and field efficiency in one water delivery schedule, it is still not impossible to identify unambiguous improvements in most water delivery schedules.

To overcome this problem, Merriam suggested intensive farmer training in field water management. Even then a second problem needs to be resolved which is the definition of the area under irrigation, as when each farmer can take water at will, conflicts are unavoidable. Surveillance of the designated command area, however, means higher transaction costs and a loss of productivity. Another criticism of Merriam's design is the high physical investment cost involved.

Changing Water Delivery Schedules: Three Cases from Baluchistan

The three case studies presented here come from a training given to selected farmers from small irrigation systems in highland Baluchistan. In this semi-arid region water is scarce and extremely precious, as it sustains a farming system of high value horticulture. The high value attached to water was the background to the participatory training in water management. In this training key farmers were asked to analyse their water delivery schedule and identify improvements in dialogue with trainer/engineers. This was preceded by the preparation of maps and an extensive investigation into the rationale of the water delivery schedule in place.

Saliaza Appozai

The first case concerns Appozai village in Zhob District, located at the tail end of the Saliaza system. The water is perennial and diverted from the Kapip Lora River at a place 20 km upstream of the village. Though Appozai is located at an altitude that allows high-value horticulture and, moreover, is at a short distance from the market town of Zhob, the command area looks forlorn, with low-value maize and underirrigated almond trees dominating the farming system.

Water Delivery System in Appozai

The poor crops reflect the inconvenient water allocation system in force in Saliaza Appozai. There are three major secondary channels in the command area. Between the three channels a time division is in place. The right channel receives water first for four days; next the central channel is served for two days; and finally the full supply goes to the left channel for six days.

An additional complication is that in the critical summer season a separate command area located close to the source is entitled to the entire Saliaza system flow for four days every 12 days, resulting in repeated four-day gaps in the. Appozai water supply. Consequently, the entitlements of the different Appozai secondary channels stretch out over 16 or 20-day periods in the summer season. This cycle is too long to enable the cultivation of important cash crops, such as tomatoes, summer vegetables and even apple trees.

A second remarkable feature of the water delivery schedule concerns the distribution of water within the branch channels. In each of the three secondary channels the flow is proportionally divided over a large number of tertiaries, in which the water flows permanently for the full duration of the secondary's turn. The size of the intakes of each tertiary is equivalent to the water rights of the command areas served by them. The base summer flow of the Saliaza channel was 75 l/s. Yet, as this is flow-divided over a large number of tertiary channels, a very large part of the flow is lost in seepage. The inconvenient irrigation cycle and the unnecessary conveyance losses in Appozai demonstrate that local resource management does not necessarily conform to the best pattern, even within the physical constraints of the system.

Several solutions to improve water availability and reduce conveyance losses were discussed with village leaders in Appozai, including the local water master. The options all revolved around reversing the order of time and flow division, namely to introduce flow division at the highest level in the system and using time shares instead of proportions of the flow to divide water over the tertiary channel. This would reduce steady state conveyance losses and make water available in each of the secondary command areas for most of the time. It would facilitate local exchange of water turns, improving flexibility and convenience.

Interestingly, the key farmers acknowledged the serious shortcomings of the present pattern and the relevance of the suggested improvements. Still, they were reluctant to correct the system, basically because they considered the costs of convincing fellow water users prohibitive. A number of reasons were given as to why it would be difficult to make innovations: the large number of water users (500), the low dependence of many shareholders on agriculture, since most landowners had ventured successfully into the construction industry, the weakness of traditional leadership and slumbering resentments between the four constituent clans of Appozai village. A final and very important obstacle was the fact that only few shareholders understood the entire system of rights and allocation rules, so that most farmers were wary of any dramatic rearrangement of the water allocation system.

Uriagai

Whereas in Appozai the lack of interest of any group to take the initiative explained the institutional stagnation, in the second case of Uriagai (Lorelai District) small groups of people, benefitting from a suboptimal status quo, blocked improvements in the water allocation system.

Uriagai is dependent on the flow of a vertical well (qanat or kareze) on the right bank of the Kohar Manda River. The water used to cross the bed of the river through an unlined farmer-built conduit. It served two separate command areas, both on the left bank. The first command area was under almond trees and high value vegetables. The entire summer flow used to be devoted to this area and did not in fact suffice to serve it adequately. The second downstream command area only received supplies in the winter season, when water supply was high and demand was low and was dedicated to low-value winter wheat. Almost all farmers owned land in both command areas.

With the help of public investments the farmer-built conduit was replaced by a concrete syphon, significantly reducing main system conveyance losses. The amount of water available to the two left bank command areas went up by 70%. The available summer base flow increased from 80 to 135 litres per second. This changed the relative scarcity of land and water in the upper command area significantly. The holding of a farmer with the fictitious name of Mohammad Siddiqui can serve as an example: before the improvement to the river crossing his 12-hour water share just about sufficed to irrigate his orchard. In his own assessment he was in fact slightly underirrigating. The increase in water supply changed this and eight hours of the augmented flow now sufficed. It created an excess flow of four hours in the summer.

To utilize this, Mohammad Siddiqui would have to take it to the lower command area, since all his land in the upper command area was under cultivation. This, however, made little sense, as most of the water would be lost in wetting the long earthen conveyance channel to this downstream command area. Instead, therefore he continued to utilize the water in the upper command area: over irrigating his own plot and giving away some of the excess flows, that still had land for expansion. Several farmers were in a similar situation to Mohammad Siddiqui. The solution would be to coordinate their excess flows. They could avoid excessive transient losses for instance by splitting the irrigation cycle in a component serving the upper command area and a component serving the lower command area. Resistance to this solution came from landlords with sufficient land in the upper command area. A change in the irrigation cycle would bring them no advantage, but instead would deprive them of the windfall extra water deliveries.

Zum Shah Murad

The village of Zum Shah Murad (Qila Saifullah District) is dependent on a highland horticulture farming system, that is comparable to Uriagai. Its source of irrigation water is a small spring (20 litres per second) in the hills overlooking the settlement. A water cycle of 15 days is applied throughout the year. An extra day is added that is 'sold' by the community to people with very small water shares. The transaction takes place at a price below the commercial value of water, in a gesture of solidarity. Apart from this, collective water share selling and purchasing of water is unusual. There are, however, intense exchanges of (parts of) water turns between farmers to meet crop irrigation intervals. While functional in this respect, the exchanges may also be expected to result in an unsystematic water schedule and concomitant high transient losses. The water schedule is, however, already chaotic without the exchanges, as principally different parts of the command area are served during a single day. This is a deliberate choice, related to the absence of an alternative source of domestic water, in contrast to Uriagai and Appozai, where piped systems were in operation. By scattering water turns throughout the command area in Zum Shah Murad, people living in small clusters in different parts of the command area are able to collect domestic water at a place close to their residence. This underlines an aspect of irrigation systems that is often ignored, but is nevertheless of relevance in water delivery schedules: the trade-off with the additional function of domestic water supply for which, as in the case of Zum Shah Murad, irrigation efficiency may be sacrificed.

An as yet unresolved issue is the impact of separating domestic and agricultural water supplies for irrigation efficiency. A variation on the chaotic schedule in Zum Shah Murad is to let the major secondary channels 'bleed' with a small flow to enable the collection of surface water. The total volume of water 'lost' in this manner may be significant, which poses the question of whether domestic water supply schemes may already be justified from the point of view of irrigation efficiency.

The example of Zum Shah Murad reminds us of the danger of overemphasizing economic and technical efficiency while ignoring the importance of non-economic objectives in local resource systems, in this case domestic water supply and local solidarity. Another irrigation system, where users took part in the participatory training on water allocation in Baluchistan, illustrates the same point. In Thal (Lorelai District), public investment in the long main channel improved the reliability of summer supplies and increased the available discharge in the command area.

The expectation of those responsible for the public investment was that this would result in an expansion of the highly profitable fruit orchards. This required, however, that the traditional practice of block rotation be abolished. Under this practice a different tract of land was cultivated every year and a lottery determined the tract of land that a particular shareholder in the system would cultivate. Obviously, block rotation and land allocation by lottery was incompatible with the cultivation of perennial crops. The water users of Thal hence faced a dilemma, but eventually made a conscious choice not to give up their traditional method of block rotation. The reason was that they feared that permanent individual rights to the land would facilitate transactions, which could set off an invasion by a neighbouring clan and undermine the relative harmony of the community at Thal.

Similarly, in Appozai and Uriagai gains in irrigation efficiency, identified as such by the water users themselves, were foregone. Instead of the requisite adjustments in the water delivery schedules, 'muddling through' prevailed. The reasons behind the lack of significant institutional change in the two systems differed and offer a view on the mechanics of indigenous institutional change, the interest of the different stakeholders in a resource system and their level of organization.

In Appozai the discussed improvements in the water distribution were Pareto-neutral. They would not have harmed anyone, but would have brought benefit to at least those farmers that wanted to adopt high value horticulture. The case of Appozai underlines the importance of the costs of institutional change. The water delivery schedule was highly specific and complex. Though each water user knew when to expect his turn from the system, very few farmers understood the Appozai water delivery schedule in its entirety.

This general lack of comprehensive system knowledge may also be called 'tradition. The Appozai water users were the captives of this incomplete knowledge. To change the water delivery schedule, this obstacle had to be overcome, yet no one in Appozai was willing to take the initiative and bear the cost of explaining the present system and the proposed modifications and convincing a multitude of co-users, probably as for no one in Appozai the additional individual benefit derived from the new schedule compensated for the efforts in convincing fellow farmers. Water shares in the system were generally small and the larger shareholders, who stood to gain most, derived most income from outside agriculture.

By contrast in Uriagai there were clear winners and losers. If the irrigation cycle had been adjusted to accommodate summer supplies to the second downstream command area, the persons with ample land in the first command area would have lost their additional supplies. The changes in other words were Pareto-biased and as both parties were equally strong, side-payments would have been necessary to overcome resistance from those that were negatively affected. Such compensations were however not acceptable. What was at stake was a windfall gain of excess water to farmers with ample land resources upstream, not a hard entitlement that could be negotiated about. In addition to complexity of conflicting interests, the cost of institutional change played a role in Uriagai as well. The water delivery schedule was highly specific and rule-oriented and few farmers understood the full system. Furthermore, there were the many inter individual adjustments and water deliveries which were hard to disentangle. Like Appozai, the water delivery schedule in Uriagai exemplifies the paradox of a specific, but immutable institution.

This raises the issue of inequity and institutional change. In case the of polarized control of power one would obviously expect that institutional change would be blocked if the party in command were to loose from the change; and would succeed if the ruling party stood to benefit.

What is evident is that the step from local economic opportunities to local institutional change, as often casually assumed to take place, is not always obvious. Non-economic considerations, conflicting interests and the question of who is going to bear the costs of the change process may stand in the way. In this respect the theory of critical mass in collective action as formulated by Marwell and Oliver (1993) is illuminating. Marwell and Oliver have postulated that one-time collective action (as required in bringing about change) is more likely to take place in the case of heterogenity. In the case of changing a water delivery schedule, this could be translated as the presence of a group of individuals with more than average interest in the improved schedule and more than average resources to bring about the change. This group would be willing to bear the initial costs of effecting the change, thus lowering the threshold for the others to follow. The consequence of this micro-social theory is that changes that bring a small benefit to all people may not necessarily materialize in contrast to changes that bring significant benefit to few.

It is here that the degree of organization comes in. Where individual users are already organized, the cost of institutional change will be lower and it will be easier to capture the marginal overall gain, or even, in the case of Pareto-biases, to negotiate side-payments. The problem with highly specific institutions is that precisely such an actively coordinating organization is missing. Water users become captives of their own schedule. In such cases it may not be wise to rely on user-driven indigenous change, but to consider outside intervention, though while respecting the sovereignty of the local users, for instance, by participatory training, awareness creation and reinforcing local organization.

Applicability of Irrigation Scheduling in Seveloping Countries

Irrigation scheduling is an important tool for optimizing the use of water resources where irrigated agriculture has a long tradition. By contrast, the operation of new irrigation districts in developing countries poses two major challenges:

- the farmer's acceptance of the larger commitments involved in irrigated agriculture (more work, care and discipline) than in rainfed agriculture;
- a suitable organization for operation and maintenance with skilled staff and adequate financing.

In the first stage all efforts should be directed towards increasing the irrigated area with an easy system. The optimization of water use through scheduling procedures should be deferred to a later stage, when the farmers have fully accepted the irrigation practices and are ready to follow more flexible rules. There is ever greater competition among the different forms of water uses: domestic and public water supply, industry, agriculture, and environment protection.

Because of the increasing difficulty in finding new resources, a strict water saving policy is an obvious must.

Agriculture makes use of three-quarters of the world's fresh water supplies. The main forms of water saving are:

- the increase of overall efficiency (reduction of conveyance, water distribution and application losses);
- the reuse of drainage water (within acceptable salinity limits);
- irrigation scheduling to make available to the biological process the correct amount of water at the appropriate time.

The three forms of water saving, separately or jointly employed, may have alternative goals:

a. to maximize the irrigable area (and thus the number of beneficiaries) for a given annual water volume;

b. to maximize the production obtainable per unit of water;

c. to maximize the average economic benefit (gross marketable product or net income) obtainable in a farm that comprises irrigated and non-irrigated land (dry farming crops, where feasible, or fallow).

The choice of the goal to be reached depends on multiple considerations of the economic, social and financial order; as a long-term policy, hypothesis (b) should be the right option. In all events, the choice will condition the procedures for conceiving and applying the water-saving technologies.

The aim of this to estimate how the irrigation scheduling processes can be applied in developing countries when new irrigation districts are to be implemented; which constraints and limitations hinder the achievement of satisfactory results; by what steps progress should be adopted. Irrigation scheduling has always been carried out in the planning stage; the recent tendency is to perform it continuously during the operation phase, as a management tool.

The choice of the irrigation volumes and of the relative application times constitutes the starting point for the design of the irrigation network. These parameters are identified in the planning phase on the basis of:

- a hypothesis of climate conditions deduced from available data;
- a hypothesis of the cropping pattern;
- a hypothesis of a sowing and crop development calendar.

Starting from these three hypotheses, the irrigation season is usually subdivided into several periods, each one being characterized by a value of unit water requirement (m³/day or l/s per hectare). The highest (peak) of these values is adopted for sizing the irrigation network.

Taking into account the hydrological characteristics of the soils, a basic constraint, and the water application procedures, the interval between two subsequent waterings is defined, where rotational delivery is adopted. Similar calculations are required for rice fields or localized irrigation.

Comparing the water requirements with the volume of the available resource (natural flow or stored volume), the irrigable area is then estimated.

According to the classical operation method, in each period the management releases into the network the discharge corresponding to the water requirement foreseen for that period. The water distribution to each user takes place with the predetermined interval and volume, for a time proportional to the acreage. The user only has the option to refuse either totally or partially the volume assigned to him. Once the system is operating, actual conditions (such as climate, cropping pattern, sowing calendar) differ from the assumed conditions in a more or less considerable and variable way; also the characteristics of the soils can differ somewhat in each single plot from those assumed for the whole district or for each subdistrict. From all this springs the wish to adapt, day by day, the amount of water supplied to the actual biological requirements. At first this wish gave rise to the setting up of 'on-demand' systems with each user having a greater or lesser degree of liberty to choose the time and volume of his withdrawal from the distribution network. The distribution network must have an operating flexibility so as to adapt itself to the variability of the withdrawal. The substitution of rotation irrigation with on-demand irrigation means that the decisions are transferred from the management to the single farmers; the management should intervene only in the case of emergency.

Past experience has shown, however, that in order to allow each single user sufficient autonomy (especially as regards the choice of the irrigation time), the network has to be oversized in an unsustainable way, with high construction and maintenance costs.

Consequently a new tendency has gradually emerged. The central management will modify, day by day, delivery volumes and times in accordance with the actual conditions and requirements. This means taking into account what is good for each single farm (for instance, with respect to the actual situation of the crops), but in an overall coordinated way for the whole district.

Irrigation scheduling has a higher efficiency when the water is derived from surface or ground storages. In the case of river diversion with pumping the benefit is given by energy-saving. The water saving is greater when irrigation occurs in variable climate conditions, with occasional but not negligible rainfalls. The techniques are more efficient in piped distribution networks and where single-crop farming prevails.

In its schematic lines the problem is as follows: given an irrigation district commanded by a storage structure, how to regulate the offtake from the reservoir and the distribution of water to the users in order to reach the selected goal. To this effect it is necessary: (i) to always be informed about the effective condition of the soil moisture; (ii) to foresee on the basis of the actual cropping pattern what the water consumption will be in the near future; and therefore (iii) to determine 'when' and 'to which extent' the soil moisture should be increased.

The concept and application of the above process involve sophisticated technology. In fact, reference is made to the construction and operation of a model in which interact:

- the climate conditions (essentially: rain, temperature, relative humidity, wind);

- the water requirements of each crop during each development stage;
- the water retention characteristics of the soils;
- the conveyance capacity of the network from the main regulation points to the single farms in order to properly regulate the water distribution and utilization.

The following are therefore necessary:

- an accurate collection and transmission (in real time) of the data from the periphery to the centre;
- an intelligent reaction at the centre for processing the data received and determining the appropriate consequences;
- a whole network of telecontrols to convey the required discharges into the various network stretches and to operate the water distribution devices to the single users or groups of users;
- a total readiness by the users to accept what is being decided by the management.

The introduction of an integral scheduling system in an established irrigation district entails a series of economic and organizational problems. However, it is not an impossible undertaking if suitably presented and explained to the farmers. The users with long irrigation experience are ready to appreciate the advantages derived from lower water consumption: lower charges, reduced energy consumption, prevention of waterlogging, less drainage requirement, less waste of fertilizers, etc.

At the same time, long irrigation practice enables the users to:

- estimate the importance and interdependence of the factors affected by the biological cycle;
- prefer the increase of net income to the achievement of exceptional unit production peaks;
- appreciate the validity of the suggested procedures, organizational patterns and highly sophisticated equipment;
- entrust the management with the control of the irrigation system.

The situation and prospects in the new irrigation districts realized in developing countries appear completely different. The emerging economic and management problems, which are referred to below, are certainly important, but the aspects and the constraints concerning the human factor are prevailing.

The introduction of irrigation constitutes an overwhelming event. First of all, because it is a collective fact that imposes a more or less tight, but nevertheless binding, discipline. Secondly, because it introduces new crops and in a period of the year in which nature 'is sleeping' and man 'is resting'. Finally, as it requires an intensity and continuity of work that rainfed farming usually does not demand.

Faced with this overwhelming situation, the novice is worried and suspicious. Little by little one has to convince him/her that irrigation is a simple practice; that to the increased work there corresponds a more than proportional income; that the network is collective but in the farm everybody is autonomous; that the results will always be better as experience becomes more thorough.

To say that this convincing action is simple and produces rapid results means not telling the truth. However, it is a necessary action that has to proceed, with absolute priority, until a real irrigation conscience has been implanted in the minds of the farmers.

In the above situation the primary goal is to start irrigated farming, i.e., to have a certain number of simple rules accepted and a certain number of operations carried out: preparation of the soil; sowing or transplanting; water application at a given interval and time (for instance Mr. XY will receive water every Monday from 9 a.m. to 4 p.m.); harvesting.

Never tell Mr. XY that because evapotranspiration is lower than usual next Monday water will only be available from 9 a.m. to 2 p.m.; or that water will not be available next Monday but next Wednesday. The farmer who has accepted a rule with difficulty does not tolerate changes to the rule and gives up the undertaking. To avoid farmers giving up the undertaking and the system remaining partially or totally unused is, therefore, the primary goal. At the beginning other objectives, even though important, must be subordinated to it.

A second issue refers to maintenance problems in developing countries. An efficient irrigation scheduling requires sophisticated equipment for teledetection, telecontrol, and data processing. Generally speaking electro-mechanical equipment frequently falls out of service and its repair often takes a number of months. Spare parts are usually unavailable in a short time.

The installation of such equipment should be avoided as far as possible. We have noticed that in some cases the water distribution scheme has been based on fixed weirs because 'also a sluice gate needs maintenance'.

In addition to the previous considerations it must be borne in mind that an irrigation system suitable for an efficient irrigation scheduling requires:

- higher initial investment;
- trained and efficient management;
- proper maintenance;
- sufficient financial flow to cover the management, operation and maintenance costs.

Many irrigation structures in developing countries have been only partially utilized or even abandoned because of organizational and financial shortcomings, but the simpler the system, the easier it is to operate.

It can thus be suggested that in developing countries new irrigation projects should:

- be implemented in such a way as to allow, in the initial stages, an extremely simple operation, even if higher than optimal unit water consumptions are required;
- be designed so as to allow further introduction (final stage) of all suitable measures and procedures to minimize water consumption through improvement of the irrigation network and full application of the irrigation scheduling procedures;
- have a command area corresponding to the reduced consumptions of the final stage, accepting that at the beginning the actually irrigated area is only a part of the command one.

The opinion can be expressed that irrigation scheduling procedures represent an important tool for optimizing the use of water in countries with a great tradition in the irrigation sector; and that, vice versa, in developing countries where irrigation is a new fact, it is better to postpone irrigation scheduling until a later consolidated phase.

This opinion is in accordance with a conclusion of a well-known Cornell University study (Studies in Irrigation No. 1-Ithaca: Cornell 1984): in the sector of irrigation projects many failures can be attributed to 'directly transfer technology that is successful in the developed countries to a developing country'.

Charging for Irrigation Water by Volume-electricity would Conserve Water Resources in Greece

Most of the area of the Greek plains, which account for 80% of the total irrigable area, is irrigated by collective irrigation networks.

The charge for water is based on the area irrigated. This practice results in water and energy losses and other secondary problems like salinity and alkalinity. This paper aims to persuade Greek farmers to accept being charged for water according to the volume or electricity consumed, as this practice saves water resources and energy, improves distribution and application efficiency, increases crop production, and decreases operation and maintenance expenses. To achieve the above-mentioned goal, the case of a collective irrigation network located in northeastern Greece has been studied. The irrigated area of this network is 7 830 ha and the payment for the irrigation water is based on the electricity consumed for pumping the irrigation water. Comparing the current practice of charging for the water by the electricity consumed with the one used some years ago of charging by area, it is clear that the land reclamation organizations all over the Greek plains, which are responsible for operating and maintaining the irrigation networks, should adopt the system of charging for irrigation water by volume or electricity.

The noticeable reduction in precipitation in the Mediterranean zone in the last ten years along with the increase in the area of land under irrigation has resulted in most of the water resources in Greece being utilized. The search for new water resources of good quality has become a difficult task. The rational management of irrigation water can reduce the losses so raising the potential for irrigating more land and for supporting the existing irrigation networks in periods of water shortage.

Irrigated agriculture, which is the main consumer of good quality water resources, can substantially reduce water losses, especially in surface irrigation where the losses exceed 50%. The main presuppositions for achieving this goal are: (i) knowledge of crop water requirements; (ii) the measurement of water distributed to the network or applied over the field; (iii) the good application over the field; and (iv) the charge for the irrigation water consumed.

This paper aims to demonstrate to Greek farmers and to the land reclamation organizations, through an existing application in north Greece, how charging for the water consumed results in water economy. For this purpose, the case of a collective irrigation network is studied.

Analysis of Existing Situation

In the Greek plains, crop water requirements are satisfied from collective irrigation networks. The management, operation and maintainance of these networks are the responsibility of the local land reclamation organizations. The total annual cost of the above-mentioned activities is allocated to each farmer served by the networks on the basis of the area cultivated and not on the water volume consumed. In the case of rice, the farmers pay more per hectare because they use more water.

The irrigation network studied is located in northeastern Greece and consists of 293 deep wells irrigating 7 830 ha (LRO, 1994; Land Reclamation Service, 1994: pers. comms.). Each well is used by a group of farmers owning neighbouring fields. The sprinkler irrigation system is semi-portable, having fully portable hand-move laterals or travelling machines and permanent buried mains. A pumping plant has been constructed at each well to provide the pressure necessary for sprinkler irrigation. The members of each group irrigate their fields whenever they want. Each well serves 20-30 ha.

Until 1981, the water charges was based on the area irrigated. This meant that the actual overall operation cost was divided by the above-mentioned area without taking into account the volume of water consumed by each farmer. As a result of this system of charging, farmers used to irrigate their fields in a way that resulted in extremely high water losses, especially from runoff. In 1982, after the Land Reclamation Organization's (LRO) proposal, the charge for water was based on the volume consumed. There were two problems: (i) the flow meter readings were strongly influenced by sand coming from the wells; and (ii) the maintenance cost of the flow meters was pretty high. So, the LRO decided to charge for the volume of water consumed on the basis of the energy needed to operate the pumping stations. More specifically, each pumping plant has a register where the energy consumed by each farmer is recorded.

The LRO writes down the readings of the electricity meter at the beginning of the irrigation period. The first farmer to irrigate his field writes down his name, community, the date of irrigation and the readings of the electricity meter at the beginning and the end of the irrigation. The following irrigators of the same group do exactly the same whenever they want to irrigate their fields. In this way, the energy consumed per irrigation is recorded and each farmer of the group checks the written entries of the farmer before. At the end of the irrigation period the LRO checks and analyses the written data and charges each farmer for the total energy consumed by him during the irrigation period. This means that the actual overall operating cost is allocated to each farmer on the bais of the energy consumed.

Methodology

Although the practice of charging for water on the basis of the energy consumed was applied all over the LRO's area, only a small part of this area (Chimonion village) had reliable data for evaluating the effect of this system of charging on the water economy from 1981 to 1994. For this period, a comparison between the 1981 data (year of charging by area) and the 1982-94 data (years of charging by energy consumed) was conducted.

To have a reliable comparison, the net irrigation requirements of the crops (In) have been calculated in Chimonion village for the 1981-94 period using the field water balance In = Etc.-(Pe + Gw + AWs), (1) where Etc. is evapotranspiration, Pe is rainfall, Gw is groundwater and AWs is the decrease in soil water that can be used during the period.

The use of combined equations (Doorenbos and Pruitt, 1977; Jensen *et al.*, 1990) to estimate reference evapotranspiration (ETo) is not applicable in this area because the available climatic data are for air temperature and rainfall only. So, the Blaney-Criddle equation (Blaney and Criddle, 1950, 1962) was used:

$$ETo = (0.46\ T + 8.13)\ p,\ (2)$$

where T is the mean daily temperature in degrees Celsius for the month under consideration, p is the mean daily percentage of total annual daytime hours obtained from tables for a given month and latitude. The Etc. was calculated using the appropriate crop coefficient (kc) for the area (Papazafiriou, 1991):

$$Etc. = ETo * kc,\ (3)$$

The crop coefficient of potatoes has not been calculated by Papazafiriou (1991), so it has been estimated on the basis of crop characteristics, sowing data, stages of crop development, length of growing season and climatic conditions (Doorenbos and Pruitt, 1977).

The mean monthly effective rainfall was estimated by the evapotranspiration/precipitation ratio method (USDA, 1967). The crop area data was obtained by the Land Reclamation Organization at Orestias (LRO, 1994).

Irrigation Scheduling in a Developing Country: Experiences from Tamil Nadu, India

Irrigation scheduling is one of the managerial activities that aim at effective and efficient utilization of water. A number of research findings and techniques related to irrigation scheduling are available today. Despite this, irrigation scheduling is only at inception level in most developing countries. This paper illustrates the status of irrigation scheduling and the constraints in the practice of scheduling at the field level from the perspective of a developing country, taking the case of Tamil Nadu in India.

The growing competition for water between the agricultural and non-agricultural sectors has increased the concern for the sustainability of the irrigated agriculture system. The need for increasing agricultural production demands an increase in the irrigated area regardless of the water resources availability for irrigation. Water has thus become a limiting factor in many of the irrigation projects. This necessitates an efficient and effective utilization of water through various water conserving methods.

Irrigation scheduling is a means of conserving water which helps in making decisions on allocation of quantity and timing of water supply commensurate with crop needs. It is one of the key activities that has the potential to improve performance of the system, especially its productivity, equity and stability (Chambers, 1983).

Research has made available a large number of findings useful for irrigation scheduling, such as those of crop water requirement estimation methods, soil moisture depletion and allied procedures, water stress indicators, water and crop yield functional relationships, and the use of remote sensing techniques. These water-saving methods and techniques are not widely practised in developing countries, particularly in the southern part of India. Irrigation systems are fascinating and complex in nature, in the sense that each of these systems has its own characteristics in its physical conditions, climatic events, design criteria, institutional arrangements, resource availability, water rights, operational rules and sociocultural practices, and these characteristics are mostly interactive in nature.

The introduction of any new procedure or technique and the degree of its success largely depend upon these system characteristics. There are wide variations between the developed and developing countries in terms of these characteristics and therefore scheduling is still at inception levels in most of the developing countries like India. This paper addresses some of the constraints in the practice of irrigation scheduling at field level, from a perspective of irrigation systems in a developing country, taking the case of Tamil Nadu in south India.

The irrigation systems in Tamil Nadu are jointly managed by the irrigation agency and the farmers. Although there exists no well defined policy or management framework for scheduling in the state of Tamil Nadu, the prevailing practices such as crop localization, monsoon dependent water allocation, and duty based design and water distribution pose constraints in the practice of scheduling. These constraints are discussed in detail.

Constraints in Practising Irrigation Scheduling

Localization Policy

The concept of localization has been in practice for a long time in India. All along, the accepted practice has been to include all cultivable land under command in the ayacut. Once the land is identified as 'commendable' under an irrigation system, it remains as such forever in the revenue records and is eligible to draw water from the system forever. No special permit is necessary season after season and thus it confers permanency in irrigation. In general, the command areas are localized as 'wet' (for water intensive crops such as paddy, sugar cane and banana) and 'irrigated dry' (for crops such as groundnut, onion and cotton), leaving decisions on crop choices to farmers. This strategy constrains scheduling in two ways.

This localization policy permits farmers to take the freedom of crop choices, even though it is authorized for a particular crop. Farmers grow one or more crops in a diversified manner according to their family needs, local expertise and market fluctuations.

The localization policy, by way of conferring rights for water along with land, gives the farmers a notion that they acquire the riparian rights on the use of water, and they try to seek legal remedies whenever changes are introduced in the interests of better water management.

Localization, therefore, gives little scope for the government agency for revising the crop policy or operational procedures. For example, the Sathanur Irrigation System in Tamil Nadu is localized for groundnut, an irrigated dry (ID) crop, in the direct command of 14 750 ha. Farmers, however, prefer to grow paddy (for family needs or due to land suitability) and sugar cane (for high returns) in addition to the ID crop. These are grown in 30-35% of the command in a scattered way. Likewise, banana (a crop of high profitability) is grown in 15% of the command area of 34 794 ha in the Tamiravaruni System.

The diversified cropping system constrains the regularization of water delivery.

Water Allocation Strategy

The strategy for water allocation and use of irrigation water in the south Indian systems has been neither demand based nor supply based, but location based (Mohanakrishnan, 1990). Seasonal and in-seasonal allocations depend on the resources available, crop seasons and priorities for water allocations prevailing in the particular location of the system. The irrigation engineers, usually the executive or assistant executive engineers, make the allocation decisions according to the 'water duty' specified at the heads of the main and branch/ distributary canals and the command area under those canals. (Water duty is locally denoted as the number of acres that can be irrigated with a cussed of water during crop period.)

In general, the crop seasons are planned to coincide with the monsoons (southwest monsoon between June and September and northeast monsoon between October and December) in most of the irrigation systems in south India. For example, in a few systems like the Tamiravaruni and the Periyar-Vaigai in Tamil Nadu, the reservoirs fill with the progress of the monsoon and cultivation starts as the monsoon proceeds. Here there are two complex situations that make crop planning and allocation decisions difficult:

 • Inflow may dwindle after an initial promise of a good monsoon:
 • The monsoon may break a little late, and the initial promise is by itself not good.

In the first case, the current practice is to reduce the supply after the initial crop period, and this might affect the crop growth. In the second case, the date of release is postponed, anticipating the strengthening of the monsoon to the extent that the crop does not become affected. The allocation decisions thus cannot be made without some risk. Even the decisions as to when the season starts and when it ends, the important decisions at the macro level are not very easy to make. Farmers also make their decisions on crop choices and planting dates in the face of uncertainty of occurrence of rainfall and the building up of storage in the reservoir. This argument emphasizes that the 'timeliness dimension' is very complex in the case of scheduling in systems in which the crop seasons are monsoon based.

Nevertheless, under these situations, while preparing the schedule, the system manager makes assumptions regarding the extent of crop area, timings of land preparation and the transplantation. These assumptions do not coincide with the actual field practices, requiring a revision in the schedule. There is no inbuilt feedback mechanism to revise the schedule on a real-time basis.

Water Distribution Strategy

In south India, irrigation systems are jointly managed by the government agency and farmers. The responsibility of water distribution up to the outlet is in the hands of the government agency, and the beneficiaries take over the responsibility below the outlet. Water is distributed at different parts of the canal network, such as branch canal and outlets, by the agency according to the duty specified. Distribution below the outlet in practice is generally to stand in queue when the lands below each turn-out are fed one after the other from head to tail. This is mostly done by mutual consent by a group of farmers. This kind of distribution practice followed by the farmers works satisfactorily as long as there is adequate waters, but it is very difficult during scarcities, leaving the tailenders to suffer.

Experience has shown that there are difficulties in the jointly managed irrigation systems. These are described below:

Systematic Scheduling: It has been noticed that in many systems the main system scheduling done by the irrigation agency and the on-farm scheduling practices adopted by the farmers do not match in terms of time of operation (Santhi, 1994). Examples of this situation can be seen in terms of the Sathanur irrigation project. The distributary canals are scheduled to be operated for a duration varying from 5 to 10 days according to the size of the command, whereas the on-farm schedule is prepared on a weekly basis (7 days) specifying timings for individual farmers (with the assistance of the Agricultural Engineering Department). This suggests that a systematic scheduling based on a bottom-up approach or top-down approach on a service agreement basis is very important for field implementation. This has been tried on a small scale in the Thindal area of the Lower Bhavani Irrigation System in Tamil Nadu through a farmers' association with the assistance of the Irrigation Community Organizers (ICOs) and it has been found to work well in the past decade. Once the water is delivered in the outlets by the irrigation agency, the farmers irrigate their farms according to the timings specified for them. The ICOs act as catalysts between the agencies involved and farmers. The ICOs assist the farmers in on-farm development works as well.

Revealed Farmer-Agency Perception: There are two aspects to the farmer-agency perception that are relevant here: one is that of adequacy and the other is of timeliness.

Adequacy: In the farmers' perception it is time and again made out that there is inadequate water to the crops, when in fact, the measurements by the agency have proven otherwise. For example, the farmers of the Nathiyunni Channel command in the Tamiravaruni system have shown that stagnating water in the paddy fields is like building up a stock of water and that when it is not possible, they have a sense of inadequacy of water for the crop grown, as during the Kar season (June-September) of 1993. The research study done in the same channel during the same season reports that the Relative Water Supply (RWS), which is a measure of adequacy, has been computed as equivalent to 1.1, that is, 110% of the water requirement (Rajan, 1993). Yet, 84% of the farmers perceived the supply as having been inadequate for their crop. Similarly, in Sathanur, the irrigation interval suggested for the ID crop of groundnut was 15 days, whereas the farmers' perception was that of a week at the most, and thus they irrigated the crop once a week.

Timeliness of Revealed Convenience: Farmers prefer to do the agricultural operations at their personal and managerial convenience depending on the availability of inputs/credit/labour. For example, the tailend farmers of the North Kodaimel Azhagian Channel of the Tamiravaruni system were able to transplant their paddy only after a month's delay in comparison with their counterparts at the head reaches during the Kar season in 1994. Because of labour shortage the transplantation could not be done. This subsequently delayed the harvest of that crop. Farmers have also shown a preference for not irrigating the fields at night, for personal convenience. It has also been seen that the farmers prefer to complete their cultural operations at a lower cost. As such, they try to use their own labour and minimize the hired labour. This might cause delay in the completion of the agricultural activities. Hence, the planned schedules could not be implemented effectively in the field.

The above explanations underline the need for service agreements between the agency and the farmers.

Infrastructure, Canals and Flow Control Systems

The design as well as the operation of the distribution network in practice in the southern region of India is based on the average 'duty'. To date, a duty of 60 acres per cussed for wet and 120 acres per cussed for irrigated dry crop are adopted for designing the distribution system. The duties adopted at different levels of the canal network in the case of the Sathanur Irrigation System are given below as an example.

Sathanur left bank canal	
Duty at the head of the main canal	1.15 l/s/ha
Duty at the head of the major distributary	1.10 l/s/ha
Duty at the head of the minor distributary	1.05 l/s/ha
Duty at the head of the outlet level	1.00 l/s/ha

Duty-based design and operational procedures do not give due consideration (a) for the 'conveyance losses' occurring in the canal network, and (b) for meeting the 'crop water requirements in time'.

The average duty considers that the lengths and conditions of the canals are the same and therefore the losses are considered to be uniform. However, they are not so in almost all of the systems in Tamil Nadu.

Delivery of water according to the duty and the crop water requirements at different stages (for groundnut, estimated at 50% efficiency) differ in quantity over time in the case of the Sathanur system. Interventions have to be made in the scheduling procedure by considering the crop water requirements and the conveyance losses occurring in the system such that they match with the existing design. This requires meteorological instruments, flow control and measuring devices and communication facilities for collection of data, elements which are not there in most of the system.

Conclusions

This paper has highlighted the following points as constraints in the practice of irrigation scheduling in the case of irrigation systems in Tamil Nadu in India, from the experience of a developing country.

This paper also suggests a few priorities for developing countries:

- The localization policy in practice for a century permits farmers to make one or more crop choices. The diversified cropping system (rice and non-rice crops) with different crops in differing growth stages and varying areal extent under the same sluice/outlet makes irrigation scheduling a difficult task in the field, considering the controls available in the system at the outlet heads and there being no measuring devices to enable the required quantum of water.

- Decisions on the start of the crop season, extent of the area allocated for irrigation, end of the season and crop planning are very difficult, as the reservoir storage and cultivation proceeds with the monsoons in most of the systems in the state.

- In the case of jointly managed systems, there are no service agreements between the agency and the farmers at present. They are important for managing the different perceptions of the farmers and the agency involved.

- The design and operation procedures of the irrigation systems in the state are duty based. Scheduling has to be done with suitable interventions, like introducing the crop water requirement concepts, allowing for losses in the network and using measuring devices.

These conclusions have been drawn from our research activities in many of the irrigation systems in Tamil Nadu. Nevertheless, irrigation scheduling is a key for improving the management of the irrigation system. Scheduling is gradually becoming recognized as a subject of importance in this country. However, not all the available scheduling techniques can be transferred. The physical, institutional, economical and sociocultural characteristics of the irrigation systems vary widely in the case of the south India, and they require their own categories of facilities, interventions and priorities.

Irrigation Management

Irrigation is the artificial *exploitation* and *distribution* of water at *project level* aiming at *application* of water at *field level* to agricultural crops in dry areas or in periods of scarce rainfall to assure or improve crop production. This chapter is about organizational forms and means of management of irrigation water at project level.

Water Management

The most important physical elements of an *irrigation project* are *land* and *water*. In accordance with the propriety relations of these elements there may be different types of water management :

- the communal type
- the enterprise type
- the utility type.

Communal Type

Until the end of the 19th century the development of irrigation projects occurred at a mild pace, reaching a total area of some 50 million ha worldwide, which is about 1/5 of the present area. The land was often private property or assigned by the village authorities to male or female farmers, but the water resources were in the hands of clans or communities who managed the water resources *cooperatively*.

Enterprise Type

The enterprise type of water management occurred under large landowners or agricultural corporations, but also in centrally controlled societies. Both the land and water resources are in one hand. Large plantations were found in colonised countries in Asia, Africa, and Latin America, but also in countries employing slave labour. It concerned mostly the large scale cultivation of commercial crops such as bananas, sugarcane and cotton. As a result of land reforms, in many countries the estates were reformed into a cooperatives in which the previous employers became members and exercised a cooperative form of land and water management.

Utility Type

Irrigation canals of the Gezira Scheme, Sudan, from space, 1997, with the utility type of management. The utility type of water management occurs in areas where the land is owned by many, but the exploitation and distribution of the water resources are managed by (government) organizations.

After 1900 governments assumed more influence over irrigation because :

- water was increasingly considered government property owing to the increasing demand for good quality water and the reducing availability
- governments embarked on large scale irrigation projects as they were considered more efficient
- the development of new irrigation schemes became technically, financially and organizationally so complicated that they fell outside the capabilities of the smaller communities
- the import and export policies of governments required the cultivation of commercial cash crops whilst, by controlling the water management, the farmers could be more easily guided to plant these kind of crops.

The water management signified a large subsidy on irrigation schemes. From 1980 the operation and maintenance of many irrigation projects was gradually handed over to water user organizations (WUA's) who were to assume these tasks and a large part of the costs, whereby the water rights of the members had to be respected. The exploitation of water resources via large storage dams-that often provided electric power as well-and diversion weirs normally remained the responsibility of the government, mainly because environmental protection and safety issues were at stake. In the past, the utility type of water management witnessed more conflicts and disturbances then the other types.

Tariffs

Irrigation water has a price by which the management costs must be covered. The following tariff (water charge) systems exist :

- No tariff, the government assumes the costs
- Tariff in labour hours, which holds mainly in communal types of management in traditional irrigation systems
- Yearly area tariff, a fixed price per ha per year
- Seasonal area tariff, a fixed price per ha per season with the higher price in the dry season
- Volumetric tariff, a fixed price per m3 of water; the consumption is measured by water meters
- Block or stepped-up pricing for water use per ha; the price increases as the water consumption per ha falls in a higher block.

The use of groundwater for irrigation is often licensed by government and the well owner may be permitted to withdraw only a maximum volume of water per year at a certain price.

Cost Recovery

The recovery of water charges may be below target, because :
- The revenues accrue to a (government) organization other than the one responsible for the management
- Farmers and water users have no say in the water management
- Lack of communication between farmers and project managers
- Poor farmers are unable to comply
- Farmers do not receive water according to need; for example insufficient quantity and/or inappropriate time
- Corruption at management level.

Cost Coverage

The cost recovery is often insufficient for full cost coverage, for example:

Country	Cost recovery (%)	Cost coverage (%)	Remarks
Argentine	67	12	low tariff: $70/ha/year
Bangladesh	3-10	<1	tariffs not enforced
Brazil, Jaiba project	66	52	
Colombia	76	52	
Turkey	76	30-40	
Sri Lanka	8	<1	tariffs not enforced

WUA's

From 1980 programs were developed to transfer the operation and maintenance tasks from the government to water user associations

(WUA's) that show some resemblance to water boards in the Netherlands, with the difference that it concerns irrigation rather than drainage and flood control. An effective development occurred in Mexico, where in 1990 a program of WUA's was initiated with tradable water rights. By 1998 some 400 WUA's were in operation commanding on average 7600 ha per WUA. They were able to recover more than 90% of the tariffs, mainly because they had to be paid in advance. Government subsidies to the water distribution and maintenance reduced to only 6%.

Water Delivery Principles

In large irrigation schemes, the distribution of irrigation water and the delivery at the farm gate is often arranged by *rotational turns* (e.g. every fortnight). The quantity of water to be received is often proportional to the farm size. As the canals usually transport constant flows, the water is being received during a period of time proportional to the farm size (e.g. every fortnight during 2 hours).

In smaller irrigation schemes the water delivery may be arranged "on demand" with water charges are on a volumetric basis. This requires a precise bookkeeping system. As the demand may be fluctuating over time, the distribution system and infrastructure is relatively expensive because it must be able to cope with periods of peak demand. During periods of water scarcity, negotiations are due to regulate the supply. From point of view of efficient irrigation water-use this is the most effective system. In projects with an uncertain supply of water due to annual variations in river discharge, water users at the top (the head users) of the irrigation system (i.e. near the system's take-off point) often have preference, to a certain extent, over users at the tail-end. Hence, the number of farmers that are able to grow an irrigated crop may vary from year to year according to the riparian water rights.

In regions with a structural water scarcity, the principle of *water duty* is often applied, whereby the duty per ha per season is only a fraction of the full irrigation need per ha (i.e. the *irrigation intensity* is less than 100%). Thus, farmers can irrigate only part of their land or irrigate their crops with a limited amount of water, whereby they may choose between crops with a high consumptive use (e.g. rice, sugarcane, most orchards) or a low consumptive use (e.g. cereals, cotton). In India, such practice is called *protective irrigation*, which aims at equal distribution of scarce means and prevention of acute famine. Owing to competition for water, the water delivery practices may deviate from the principles.

Water Delivery Practices

In practice the distribution of irrigation water is subject to competition. Influential farmers may be able to acquire more water than they are entitled to. Water users at the upstream part of the irrigation system can more easily intercept extra water than the tail-ender. The degree of farmers' influence is correlated to the relative position of their land in the topography of the scheme.

Tail-end Problems

R.Chambers cites authors who have reported tail-end problems. Examples are:

- The old Sardar canal project in the state of Gujarat, India, was designed with an irrigation intensity of 32%, but at the upstream part the delivery was at an intensity of 42% (i.e. 131% of the design norm) and at the downstream end it was only 19% (i.e. 59% of the norm), although the project aimed at protective irrigation with equal rights for all.

- The Sardar Sahayak Pariyojana irrigation project, an extension of the Sardar canal project with 1.7 million ha, the head farmers received 5 times more water than the tail-enders, although the project was designed for equal distribution of the scarce water.

- The Ghatampur distributary canal in the Ramganga irrigation project in the state of Uttar Pradesh, India, delivered an amount of water equal to 155% of the design discharge to the Kisarwal district canal near the head of the distributary and only 22% to the Bairampur district canal at the downstream end.

Environmental Impact of Irrigation

Environmental impacts of irrigation are the changes in quantity and quality of soil and water as a result of irrigation and the ensuing effects on natural and social conditions at the tail-end and downstream of the irrigation scheme. The impacts stem from the changed hydrological conditions owing to the installation and operation of the scheme. An irrigation scheme often draws water from the river and distributes it over the irrigated area. As a hydrological result it is found that:

- the downstream river discharge is reduced
- the evaporation in the scheme is increased

- the goundwater recharge in the scheme is increased
- the level of the water table rises
- the drainage flow is increased.

These may be Called Direct Effects

The effects thereof on soil and water quality are indirect and complex, Water logging and soil salination are part of these, whereas the subsequent impacts on natural, ecological and socio-enonomic conditions is very intricate.

Irrigation can also be done extracting groundwater by (tube) wells. As a hydrological result it is found that the level of the water descends. The effects may be water mining, land/soil subsidence, and, along the coast, saltwater intrusion.

Irrigation projects can have large benefits, but the negative side effects are often overlooked.

Reduced Downstream River Discharge

The reduced downstream river discharge may cause:

- reduced downstream flooding
- disappearance of ecologically and economically important wetlands or flood forests
- reduced availability of industrial, municipal, household, and drinking water
- reduced shipping routes. Water withdrawal poses a serious threat to the Ganges. In India, barrages control all of the tributaries to the Ganges and divert roughly 60 percent of river flow to irrigation
- reduced fishing opportunities. The Indus River in Pakistan faces scarcity due to over-extraction of water for agriculture. The Indus is inhabited by 25 amphibian species and 147 fish species of which 22 are found nowhere else in the world. It harbours the endangered Indus River dolphin, one of the world's rarest mammals. Fish populations, the main source of protein and overall life support systems for many communities, are also being threatened
- reduced discharge into the sea, which may have various consequences like coastal erosion (e.g. in Ghana) and salt water intrusion in delta's and estuaries. Current water withdrawal from the river Nile for irrigation is so high that, despite its size, in dry periods the river does not reach the sea. The Aral

sea has suffered an "environmental catastrophe" due to the interception of river water for irrigation purposes.

Increased Groundwater Recharge, Waterlogging, Soil Salinity

This illustrates an environmental impact of upstream irrigation developments causing an increased flow of groundwater to this lower lying area leading to the adverse conditions. The increased groundwater recharge stems from the unavoidable deep percolation losses occurring in the irrigation scheme. The lower the irrigation efficiency, the higher the losses. Although fairly high irrigation efficiencies of 70% or more (i.e. losses of 30% or less) can be obtained with sophisticated techniques like sprinkler irrigation and drip irrigation, or by precision land levelling for surface irrigation, in practice the losses are commonly in the order of 40 to 60%. This may cause:

- rising water tables,
- increased storage of groundwater that may be used for irrigation, municipal, household and drinking water by pumping from wells,
- waterlogging and drainage problems in villages, agricultural lands, and along roads with mostly negative consequences. The increased level of the water table can lead to reduced agricultural production.
- shallow water tables are a sign that the aquifer is unable to cope with the groundwater recharge stemming from the deep percolation losses,
- where water tables are shallow, the irrigation applications are reduced. As a result, the soil is no longer leached and soil salinity problems develop,
- stagnant water tables at the soil surface are known to increase the incidence of water borne diseases like malaria, filariasis, yellow fever, dengue, and schistosomiasis (Bilharzia) in many areas. Health costs, appraisals of health impacts and mitigation measures are rarely part of irrigation projects, if at all.
- to mitigate the adverse effects of shallow water tables and soil salinization, some form of watertable control, soil salinity control, drainage and drainage system is needed.
- As drainage water moves through the soil profile it may dissolve nutrients (either fertilizer based or naturally occurring) such as nitrates, leading to a built up of those nutrients in the ground water aquifer. High nitrate levels in drinking water

can be harmful to humans particularly for infants under 6 months where it is linked to 'blue-baby syndrome'.

Case Studies:

1. In India 2.189.400 ha have been reported to suffer from waterlogging in irrigation canal commands. Also 3.469.100 ha were reported to be seriously salt affected here

2. In the Indus Plains in Pakistan, more than 2 million hectares of land is waterlogged. The soil of 13.6 million hectares within the Gross Command Area was surveyed, which revealed that 3.1 million hectares (23%) was saline. 23% of this was in Sindh and 13% in the Punjab. More than 3 million ha of water-logged lands have been provided with tube-wells and drains at the cost of billions of rupees, but the reclamation objectives were only partially achieved. The Asian Development Bank (ADB) states that 38% of the irrigated area is now waterlogged and 14% of the surface is too saline for use

3. In the Nile delta of Egypt, drainage is being installed in millions of hectares to combat the water-logging resulting from the introduction of massive perennial irrigation after completion of the High Dam at Assuan

4. In Mexico, 15% of the 3.000.000 ha if irrigable land is salinized and 10% is waterlogged

5. In Peru some 300.000 ha of the 1.050.000 ha of irrigable land suffers from this problem.

6. Estimates indicate that roughly one-third of the irrigated land in the major irrigation countries is already badly affected by salinity or is expected to become so in the near future. Present estimates for Israel are 13% of the irrigated land,, Australia 20%, China 15%, Irak 50%, Egypt 30%. Irrigation-induced salinity occurs in large and small irrigation systems alike

7. FAO has estimated that by 1990 about 52×10^6 ha of irrigated land will need to have improved drainage systems installed, much of it subsurface drainage to control salinity.

Reduced Downstream Drainage and Groundwater Quality:

- The downstream drainage water quality may deteriorate owing to leaching of salts, nutrients, herbicides and pesticides. This may negatively affect the health of the population at the tail-end and downstream of the irrigation scheme, as well as the ecological balance. The Aral sea, for example, is seriously polluted by drainage water.

- The downstream quality of the groundwater may deteriorate in a similar way as the downstream drainage water and have similar consequences.

Reduced Downstream River Water Quality

Owing to drainage of surface and groundwater in the project area, which waters may be salinized and polluted by agricultural chemicals like biocides and fertilizers, the quality of the river water below the project area can deteriorate, which makes it less fit for industrial, municipal and household use. It may lead to reduced public health. Polluted river water entering the sea may adversely affect the ecology along the sea shore.

Affected Downstream Water Users

Downstream water users often have no legal water rights and may fall victim of the development of irrigation.

Pastoralists and nomadic tribes may find their land and water resources blocked by new irrigation developments without having a legal recourse.

Flood-recession cropping may be seriously affected by the upstream interception of river water for irrigation purposes.

- In Baluchistan, Pakistan, the development of new small-scale irrigation projects depleted the water resources of nomadic tribes travelling annually between Baluchistan and Gujarat or Rajasthan, India
- After the closure of the Kainji dam, Nigeria, 50 to 70 per cent of the downstream area of flood-recession cropping was lost.

Lost Land use Opportunities

Irrigation projects may reduce the fishing opportunities of the original population and the grazing opportunities for cattle. The livestock pressure on the remaining lands may increase considerably, because the ousted traditional pastoralist tribes will have to find their subsistence and existence elsewhere, overgrazing may increase, followed by serious erosion and the loss of natural resources. The Manatali reservoir formed by the Manantali dam in Mali intersects the migration routes of nomadic pastoralists and destroyed 43000 ha of savannah, probably leading to overgrazing and erosion elsewhere. Further, the reservoir destroyed 120 km^2 of forest. The depletion of groundwater aquifers, which is caused by the suppression of the seasonal flood cycle, is damaging the forests downstream of the dam.

Groundwater Mining with Wells, Land Subsidence

When more groundwater is pumped from wells than replenished, storage of water in the aquifer is being mined. Irrigation from groundwater is no longer sustainable then. The result can be abandoning of irrigated agriculture.

The hundreds of tubewells installed in the state of Uttar Pradesh, India, with World Bank funding have operating periods of 1.4 to 4.7 hours/day, whereas the were designed to operate 16 hours/day.

In Baluchistan, Pakistan, the development of tubewell irrigation projects was at the expense of the traditional qanat or karez users.

Groundwater-related subsidence of the land due to mining occurred in the USA at a rate of 1m for each 13m that the watertable was lowered.

Homes at Greens Bayou near Houston, Texas, where 5 to 7 feet of subsidence has occurred, were flooded during a storm in June 1989 as shown in the picture.

Irrigation in Viticulture

The role of irrigation in viticulture is considered both controversial and essential to wine production. In the physiology of the grapevine, water is a vital component to function of the vine with its presence or lack impacting photosynthesis, new plant shoot growth, as well as the development of grape berries. While climate and humidity play important roles, the typical grape vine needs an average of 27 inches (680 millimetres) of water a year, ideally occurring during the winter and spring months of the growing season. In many Old World wine regions, natural rainfall is considered the only source for water that will still allow the vineyard to maintain its *terroir* characteristics. The practice of irrigation is viewed by some critics as unduly manipulative with the potential for detrimental wine quality due to high yields that can be artificially increased with irrigation. It has been historically banned by the European Union's wine laws, though in recent years individual countries (such as Spain) have been loosening their regulations and France's wine governing body, the *Institut National des Appellations d'Origine* (INAO), has also been reviewing the issue.

In very dry climates that receive little rainfall, irrigation is considered essential to any viticultural prospects. Many New World wine regions such as Australia and California regularly practice irrigation in areas that couldn't otherwise support viticulture. Advances and research in these wine regions (as well as some Old World wine regions such as Israel), have shown that potential wine quality could increase in areas where irrigation is kept to a minimum and managed. The main principle behind this is controlled water stress where the vine receives sufficient water during the budding and flowering period that is then scaled back during the ripening period where the vine then responds by funneling more its limited resources into developing the grape clusters instead of excess foliage. If the vine receives too much water stress, then photosynthesis and other important process could be impacted with vine essentially shutting down. The availability of irrigation means that if drought conditions emerge, sufficient water can be provided for the plant so that the balance between water stress and development is kept to optimal levels.

History

The practice of irrigation has had a long history in wine production. Archaeologists describe it as one of the oldest practices in viticulture, with irrigation canals discovered near vineyard sites in Armenia and Egypt dating back more than 2600 years. Irrigation was already widely practice for other agricultural crops since around 5000 BC. It is possible that the knowledge of irrigation helped viticulture spread from these areas to other regions due to the potential for the grapevine to grow in soils too infertile to support other food crops. A somewhat hardy plant, the grapevine largest need is for sufficient sunshine and is able to flourish with minimum needs of water and nutrients. In areas where its water needs are unfulfilled, the availability of irrigation meant that viticulture could still be supported.

In the 20th century, the growing wine industries of California, Australia and Israel were greatly enhanced by advances in irrigation. With the development of more cost efficient and less labour intensive ways of watering the vines, vast tracks of very sunny but dry lands were able to be converted into growing wine regions. The ability to control the precise amount of water each vine received, allowed producers in these New World wine regions to develop styles of wines that could be fairly consistent each year regardless of normal vintage variation. This created a stark contrast to the Old World wine regions of Europe where vintage variation, including rainfall, had a pronounced effect on the potential wine style each year. Continuing research explored the way that controlled (or supplemental) irrigation could be used to increase potential wine quality by influence how the grapevine responds to its environment and funnels resources into developing the sugars, acids and phenolic compounds that contribute to a wine's quality. This research lead to the development of ways to measure the amount water retention in the soil to where individual irrigation regimes could be plotted for each vineyard that maximized the benefits of water management.

Role of Water in Viticulture

The presence of water is essential for the survival of all plant life. In a grapevine, water acts as a universal solvent for many of the nutrients and minerals needed to carry out important physiological functions-which the vine receives by absorbing the nutrient-containing water from the soil. In the absence of water in the soil, the root system of the vine may have difficulties absorbing these nutrients. Within the structure of the plant itself, water acts as a transport within the xylem bring these nutrients to all ends of the plant. During the process of photosynthesis, water molecules combine with carbon derived from carbon dioxide to form glucose which is the primary energy source of the vine as well as oxygen as a by-product.

In addition to its use in photosynthesis, a vine's water supply is also depleted by the processes of evaporation and transpiration. In evaporation, heat (aided by wind and sunlight) causes water moisture in the soil to evaporate and escape as vapor molecules. This process is inversely related to humidity with evaporation often taking place at faster rates in areas with low relative humidity. In transpiration, this evaporation of water occurs directly in the wine as water is released from the plant through the stomata that is located underneath the leafs of a grape vine.

This process helps the vine combat against the effects of heat stress which can severely damage the physiological functions of the vine (somewhat similar to how perspiration works with humans and animals). The presence of adequate water in the vines can help keep the internal temperature of a the leaf only a few degrees above the temperature of the surrounding air. However, if water is severely lacking then that internal temperature could jump nearly 18 °F (10 °C) warmer than the surrounding air which leads the vine to develop heat stress. The duel effects of evaporation and transpiration is called evapotranspiration. A typical vineyard in a hot, dry climate can lose as much as 40,000 litres (11,000 U.S. gal; 8,800 imp gal) of water through evapotranspiration during the growing season. Even in a cooler, dry climate, a vineyard could still lose enough water to flood the vineyard with 16 inches (400 mm) if all that water was suddenly replaced at once.

Factors Influencing Irrigation

Climates with low humidity promote faster rates evapotranspiration which reduce the grapevine's water supply. These areas may need to utilize supplemental irrigation.

There are essentially two main types irrigation-primary irrigation, which is needed for areas (such as very dry climates) that lack sufficient rainfall for viticulture to even exist, and supplemental irrigation where irrigation is used to "fill in the gaps" of natural rainfall to bring water levels to more optimal numbers as well as to serve as a preventive measure in case of seasonal drought conditions. In both cases, both the climate and the vineyard soils of the region will play an instrumental role in irrigation's use and effectiveness.

Impact of Different Climate Types

Viticulture is most commonly found in Mediterranean, continental and maritime climates with each unique climate providing its own challenges in providing sufficient water at critical times during the growing season. In Mediterranean climates irrigation is usually needed during the very dry periods of the summer ripening stages where drought can be a persistent threat. The level of humidity in a particular macroclimate will dictate exactly how much irrigation is needed with high levels of evapotranspiration more commonly occurring in Mediterranean climates that have low levels of humidity such as part of Chile and the Cape Province of South Africa. In these low humidity regions, primary irrigation maybe needed but in many Mediterranean climates the irrigation is usually supplemental.

Continental climates are usually seen in areas further inland from the coastal influences of oceans and large bodies of water. The difference from the average mean temperature of its coldest and hottest months can be quite significant with moderate precipitation that usually occurs in the winter and early spring. Depending on the water retaining ability of the soil the grapevine may receive enough water during this period to last throughout the growing season with little if any irrigation needed. For soils with poor water retention, the dry summer months may require some supplemental irrigation. Examples of continental climates that use supplemental irrigation include the Columbia Valley of Washington State and the Mendoza wine region of Argentina.

Maritime climates tend to fall between Mediterranean and continental climates with a moderate climate that is tempered by the effects of a large body of water nearby. As with Mediterranean climates, the humidity of the particular macroclimate will play a significant role in determining how much irrigation is needed. In most cases irrigation, if it is used at all, will only be supplemental in years where drought may be an issue. Many maritime regions, such as Rias Baixas in Galicia, Bordeaux and the Willamette Valley in Oregon, suffer from the diametric problem of having too much rain during the growing season.

Impact of Different Soil Types

Soil can have a significant impact on the potential quality of wine. While geologist and viticulturist are not exactly sure what type of immutable or *terroir* based qualities that soil can impart on wine, there is near universal agreement that a soil's water retention and drainage abilities play a primary role. Water retention refers to the soil's ability to hold water.

The term "field capacity" is used to describe the maximum amount of water that deeply moisten soil will retain after normal drainage. Drainage is the ability of water to move freely throughout the soil. The ideal circumstance is soil that can retain sufficient amount of water for the grapevine but drains well enough to where the soil doesn't become water-logged. Soil that doesn't retain water well encourages the vine to easily sleep into water stress while soil that doesn't drain well runs of the risk of water-logged roots being attacked by microbial agents that consume all the soil nutrients and end up starving the vine.

The depth, texture and composition of soils can influence its water retaining and draining ability. Soils containing large amounts of organic material tend to have the highest water retention abilities. These types of soils include deep loams, silty soils like what is typically found on the fertile valley floors such as in the California's Napa Valley. Clay particles have the potential to remain in colloidal suspension for long periods of time when it is dissolved in water. This gives clay-based soils the potential to retain significant amount of water such as the clay soils of the Right bank Bordeaux region of Pomerol. Many regions with these types of water retaining soils have little need for irrigation, or if they do it is usually supplemental during periods of drought. Soils with poor water retention include sand and alluvial gravel based soils such as those found in the Barolo and Barbaresco zones of Italy or in many areas of South Australia. Depending on the climate and amount of natural rainfall, areas with poor water retention may need irrigation.

Just as having too little water is detrimental to the grapevine, so too is having too much. When vines become water-logged they become a target for various microbial agents such as bacteria and fungi that compete with the vine for nutrients in the soil. Additionally excessively moist soil is poor conductor of valuable heat radiating from the ground. In general wet soils are cold soils which can be especially problematic during the flowering causing poor berry set that could lead to coulure. It also becomes an issue during the ripening stage when vines in cool-climate regions may need additional heat radiated from the ground in order to sufficiently ripen its fruit (an example of this is the slate-based vineyards of the Mosel in Germany). Therefore, well draining soils are considered very conducive to producing quality wine. In general light-textured (such as sand and gravel) and stony soils tend to drain well. Soils heavy soils and those with high proportions of organic matter also have the potential to drain well if they having a crumbling texture and structure. This texture relates to the friability of the soil which can come from earthworms and other organisms that have burrowed tunnels throughout the soil. Much like rocks, these tunnels give water a freer passageway through soil and contributes to its drainage.

Measuring Soil Moisture

Tensiometers can be used to measure soil moisture. The components of this example include (1) porous cup, (2) water-filled tube (3) sensor-head and a (4) pressure sensor.

Because of the problems associated with water-logged and wet soils, it is important for viticulturist to know how much water is currently in the soil before deciding if and how much to irrigate. There are several methods of evaluating soil moisture. The most basic is simple observation and feeling of the soil, however this has its limitations since the subsoil may be moist while the surface soil appears dry. More specific measurements can be attained by using tensiometers which evaluates surface tension of water extracted from the soil. The presence of water in the soil can be measured by neutron moisture meters that utilize an aluminum tube with an internal neutron source that detect the subtle change between the water in the soil. Similarly, gypsum block placed throughout the vineyard contain an electrode that can be used to detect the electrical resistance that occurs as the soil dries and water is released by evaporation. Since the 1990s there has been greater research into tools utilizing time-domain reflectivity and capacitance probes. In addition to monitoring for excessive moisture, viticulturists also keep an eye for signs of water stress due to severe lack of water.

Irrigation Systems

There are several methods of irrigation that can be used in viticulture depending on the amount of control and water management desired. Historically, surface irrigation was the most common means using the gravity of a slope to release a flood of water across the vineyard. In the early history of the Chilean wine industry, flood irrigation was widely practiced in the vineyards using melted snow from the Andes Mountains channelled down to the valleys below. This method provided very little control and often had the adverse effect of over-watering the vine. An adoption of method was the furrow irrigation system used in Argentina where small channels ran through the vineyard providing irrigation. This provide slightly more control since the initial amount of water entering the channels could be regulated, however the amount that each vine received was sporadic.

Sprinkler irrigation involves the installation of a series of sprinkler units throughout the vineyard, often spaced as several rows about 65 feet (20 m) apart. The sprinklers can be set on a electronic timer and release predetermined amount of water for a set period of time. While this provides more control and uses less water than flood irrigation, like furrow irrigation the amount that each individual wine receives can be sporadic. The irrigation system that provides the most control over water management, though conversely the most expensive to install, is drip irrigation. This system involved long plastic water supply lines that run down each row of vines in the vineyard with each individual grape vine having its own individual dripper. With this system, a viticulturist can control the precise amount of water that each grapevine gets down to the drop. An adaption of this system, potentially useful in areas where irrigation may be banned, is underground sub-irrigation where precise measurements of water is delivered directly to the root system.

When and how much?

Water is very crucial during the early budding and flowering stages but after fruit set *(pictured)*, the amount of water given to the vine may be scaled back in order to promote water stress.

With abundant water, a grapevine will produce shallow root systems and vigorous growths of new plant shoots. This can contribute to a large, leafy canopy and high yields of large grape berry clusters that may not be sufficiently or physiologically ripe. With insufficient water, many of the vine's important physiological structures, including photosynthesis that contributes to the development of sugars and phenolic compounds in the grape, can shut down. The key to irrigation is to provide just enough water for the plant to continuing function without encouraging vigorous growth of new shoots and shallow roots. The exact amount of water will depend on a variety of factors including how much natural rainfall can be expected as well as the water retaining and drainage properties of the soil.

Water is very crucial during the early budding and flowering stages of the growing season. In areas where there is not sufficient rainfall, irrigation may be needed during this time in the spring. After fruit set, the water needs for the vine drop and irrigation is often withheld till the period of *veraison* when the grapes begin to change colour. This period of "water stress" encourages the vine to concentrate its limited resources into lower yields of smaller berries creating a favourable skin to juice ratio that is often desirable in quality wine production. The benefits or disadvantageous of irrigation during the ripening period itself is a matter of debate and continuing research in the wine growing community. The only area of mostly agreement is the disadvantages of water close to harvest after a prolong dry period. Grapevines that has been subjugated to prolong water stress have a tendency to rapidly absorb large amounts of water if its provided. This will dramatically swell the berries, potentially causing to them crack or burst which will make the prone to various grape diseases. Even if the berries do not crack or burst, the rapid swelling of water will cause a reduce concentration in sugars and phenolic compounds in the grape producing wines with diluted flavours and aromas.

Water Stress

One of the goals of controlled, mild water stress is to discourage the formation of excess new plant growths *(a bud pictured)* which will compete with the developing grape clusters for the vine's limited resources.

The term water stress describes the physiological states that grapevines experience when they are deprived of water. When a grapevine goes into water stress one of its first functions is to reduce the growth of new plant shoots which compete with the grape clusters for nutrients and resources. The lack of water also keeps the individual grape berries down to a smaller size which increase its skin to juice ratio. As the skin is filled with color phenolase, tannin and aroma compounds, the increase skin to juice ratio is desirable for the potential added complexity the wine may have. While there is disagreement over exactly how much water stress if beneficial in development grapes for quality wine production, most viticulturist agree that some water stress can be beneficial. The grapevines in many Mediterranean climates such as Tuscany in Italy and the Rhone Valley in France experience natural water stress due to the reduce rainfall that occurs during the summer growing season.

At the far extreme is severe water stress which can have detrimental effects on both the vine and on potential wine quality. To conserve water, a vine will try to conserve water by limiting its lose through transpiration. The plant hormone abscisic acid triggers the stomata on the underside of the plant leaf to stay close in order to reduce the amount of water that is evaporated. While conserving water this also has the consequences of limiting the intake of carbon dioxide needed to sustain photosynthesis.

If the vine is continually stressed it will keeps it stomata closed for longer and longer periods of time which can eventually cause photosynthesis to stop all together. When a vine has been so deprived of water it can exceed what is known as its permanent wilting point. At this point, the vine can become permanently damaged beyond recovery even if later watered. Viticulturists will carefully watch the plant for signs of severe water stress. Some of the symptoms include:

- Flaccid and wilting tendrils
- (During Flowering) Flower clusters that are dried out
- Wilting of young grape leaves followed by maturer leaves
- Chlorosis signalling that photosynthesis has stopped
- Necrosis of dying leaf tissue which leads to premature leaf fall
- Finally, the grape berries themselves start to shrivel and fall off the vine.

The effectiveness of water stress is an area of continuing research in viticulture. Of particular focus is the connection between yield size and the potential benefits of water stress. Since the act of stressing the vine does contribute to reduce photosynthesis-and by extension, reduce ripening since the sugars produced by photosynthesis is needed for grape development-it is possible that a stressed vine with high yields will only produce lots of under ripe grapes. Another interest of study is the potential impact on white grape varieties with enologists and viticulturists such as Cornelius Van Leeuwen and Catherine Peyrot Des Gachons contending that white grape varieties lose some of their aromatic qualities when subjugated to even mild forms of water stress.

Partial Rootzone Drying

In partial rootzone drying, half the roots are allowed to dehydrate which sends signals to the vine that is experiencing "water stress". Meanwhile the irrigated roots on the other side of vine continue to provide sufficient amounts of water so that vital functions like photosynthesis does not cease.

One irrigation technique known as partial rootzone drying (or PRD) involves "tricking" the grapevine into thinking it is undergoing water stress when it is actually receiving sufficient water supply. This is accomplished by alternating drip irrigation to where only one side of the grapevine receives water. The roots on the dry side of the vine produce abscisic acid that triggers some of the vine's physiological responses to water stress-reduced shoot growth, smaller berries size, etc. But because the vine is still receiving water on the other side the stress doesn't become so severe to where vital functions such as photosynthesis is compromised.

Criticism and Environmental Issues

The practice of irrigation has it share of criticism and environmental concerns. In many European wine regions the practice is banned under the belief that irrigation can be detrimental to quality wine production. However, in the early 21st century some European countries have relaxed their irrigation laws or re-evaluated the issue. Of the criticisms levelled towards irrigation, the most common is that it disrupts the natural expression of *terroir* in the land as well as the unique characteristics that comes with vintage variation. In regions that do not practice irrigation, the quality and styles of wines can be dramatically different from vintage to vintage depending on weather conditions and rainfall. Irrigation's contribution to the broader globalization of wine is criticized as promoting a homogenization or "standardization" of wine.

Other criticisms centre around the broader environmental impact of irrigation on both the ecosystem around the vineyard as well as the added strain on global water resources. While advances in drip irrigation has reduced the amount of waste water produced by irrigation, the irrigation of substantial tracts of land in areas like the

San Joaquin Valley in California and the Murray-Darling Basin of Australia requires massive amounts of water from dwindling supplies. In Australia, the centuries old practice of flood irrigation used in places like the Murrumbidgee Irrigation Area caused severe environmental damages from water-logging, increase salination and raising the water tables. In 2000, the Australian government invested over 3.6 million AUD into research on how to minimize the damage caused by extensive irrigation. In 2007, concerns about ecological damage to the Russian River caused government officials in California to take similar measures to cut back water supplies and promote more efficient irrigation practices.

Other uses for Irrigation Systems

In addition to providing water for plant growth and development, irrigation systems can also be used for alternative purposes. One of the most common is the dual application of fertilizer with water in a process known as fertigation. Commonly used in drip irrigation systems, this method allows similarly regulate control over how precisely how much fertilizer and nutrients that each vine receives. Another alternative use for sprinkler irrigation systems can occur during the threat of winter or spring time frost. When temperature drop below 32 °F (0 °C), the vine is at risk of developing frost damage that could not only ruining the upcoming years harvest but also kill the vine. One preventive measure against frost damage is to use the sprinkler irrigation system to coat the vines with a protective layer of water that freezes into ice. This layer of ice serves as insulation keeping the internal temperature of the vine from dropping below the freezing mark.

Tidal Irrigation

Tidal irrigation is the subsurface irrigation of levee soils in coastal plains with river water under tidal influence. It is applied in (semi) arid zones at the mouth of a large river estuary or delta where a considerable tidal range (some 2 m) is present. The river discharge must be large enough to guarantee a sufficient flow of fresh water into the sea so that no salt water intrusion occurs in the river mouth.

The irrigation is effectuated by digging *tidal canals* from the river shore into the main land that will guide the river water inland at high tide. For the irrigation to be effective the soil must have a high infiltration capacity to permit the entry of sufficient water in the soil to cover the evapotranspiration demand of the crop. At low tide, the canals and the soil drain out again, which promotes the aeration of the soil.

Irrigation of Alluvial Fans

Irrigation of alluvial fans is the use of water resources, mainly river floods and groundwater recharged by infiltration of river water, to enhance the production of agricultural crops.

• Alluvial fans, when large and flat, are also called *inland delta's*.

General Description

Alluvial fans, also called inland deltas, occur at the foot of mountain ranges and mark the presence of river floods. The rivers flowing at high speed in the mountains carry sediments. Upon losing its speed in the flat land at the foot of the mountain, the water deposits its sediments forming a cone shaped earth body. The course sediments like gravel and sand are deposited first, close to the entrance of the river into the plain. The finer sediments, consisting of silt and clay, are deposited towards the base of the cone.

Upon entering the plain, and forced by the deposits of the sediments, the river divides itself into numerous branches fanning out towards the plain. The alluvial fans contain considerable groundwater reservoirs that are replenished each year by infiltration of the water from the river branches into the usually permeable underground thus obtaining rich aquifers.

The mountainous areas usually receive more rainfall than the plains: they form a watershed and provide a source of water. In (semi) arid regions, therefore, alluvial fans are often used for irrigation of agricultural crops. The fans reveal much greenery in the harsh desert-like environment.

The irrigation methods in alluvial fans differ according to the hydrological regime of the river, the shape of the fan, and the natural resources available to maintain human life.

Types of Fans

The following alluvial fans will be reviewed in increasing order of water yield:

• The alluvial fans along the river plains near Khuzdar, Baluchistan, Pakistan. These fans are fed by small water catchments in areas of relatively low mountains. The fans are relatively small, steep, and subject to flash floods

- The alluvial fan of Garmsar, east of Tehran, Iran. This fairly large fan is fed by the *Hableh Rud* (river) with an important catchment area in the high Alburz mountain range. The river carries a large amount of water during the rainy season, otherwise the discharge is low.
- The alluvial fan of Punata in the *Valle Alto*, east of Cochabamba, Bolivia. The fan is fed by The *Rio Paracaya* river with a higher average discharge than the *Hableh Rud* river. Consequently the fan is fairly flat.
- The Okavango inland delta, near Maun, Botswana.

The delta receives an enormous amount of river inflow from Angola. Hence, the fan is so large and flat that it is rather called a delta. It takes six months for the peak inflow at the apex to reach the base of the delta.

Case Studies

Khuzdar

Average annual rainfall in Baluchistan varies between 200 and 400 mm, depending on altitude, and the main part occurs in winter (November to March). Of old, in sloping lands, farmers constructed bunds along the contour lines to capture the surface runoff. This method of water harvesting (locally called *khuskaba*) provided extra water for agricultural crops planted just up-slope of the bund, where the captured water would infiltrate into the soil and provide extra soil moisture to supplement the scarce rainfall.

In alluvial fans, the spate floods provided and additional source of water. The floods, diverted from the watercourses, were retained behind similar bunds employed in the *khuskaba* system. The method of flood interception is locally called *sailaba*. The system is combined with the tapping of groundwater from the aquifer by means of dug underground galleries, called *karez* or qanat. The *karezes* make permanent agriculture possible.

Although the *sailaba*-and-*qanat* systems cover about 20% of the agricultural land, their production is more than 40% of the total.

It is a modern development to sink deep wells in the aquifer of the alluvial fan to exploit the groundwater more effectively than do the traditional *karezes*. The owners of the wells may be entrepreneurs from elsewhere and the original population runs the risk of loosing the *karez* water when the wells draw the water table down to a deeper depth than that of the *karezes* so that the these fall dry.

Garmsar

The irrigation system for the Garmsar alluvial fan is quite well developed, to the extent that lined canals have been made and a large belt-canal crosses the fan through its middle.

Roughly, the cropped area occupies 30% of the land each season, while 70% is left fallow. The winter crops are mainly wheat and barley, while the summer crops are cotton and melons. However, the planting of the new crops is done before harvesting the previous crops. Thus, there is a period of overlap during which 60% of the land is under crops. The fallow land is continuously rotated throughout the years, so that there exists no permanent fallow land, except along the fringes at the base of the fan where soil salinization occurs.

Cumulative frequency distribution of the annual average river discharge, showing a large variation.

An estimated average annual water balance. It is seen that the storage of irrigation losses in the aquifer plays an important role. In the dry season the groundwater is used for irrigation by pumping from deep wells. A cross-section of the groundwater situation.

The water rights are expressed in *sang*, a measure of continuous flow of about 10 l/s, but in practice it varies from 10 to more than 15 l/s. The water is delivered to about 100 tertiary units (often a village), within which the water is distributed by 12-day rotations amongst the farmers who each are entitled to receive the authorized *sangs* for a fixed number of hours during each rotation period. The village communities are, at the same time, water-user associations who take care of the water-distribution within the tertiary unit and they maintain the tertiary canals.

At present, the distribution of surface irrigation water to the villages is determined by the Garmsar Water Authority on the basis of the water rights and verbal agreements and communications with the water users in the absence of a written manual. The authority also maintains the irrigation canals and structures. The structures are sometimes re-designed to adjust them to verbally communicated needs. The fair distribution of the irrigation water is not an easy job as the average annual river discharge is quite variable in the range of 5 to 20 m3/s.

The deep tube-wells are privately owned. The drilling of wells is subject to license. Recently, the licensing has stopped for fear of over-exploitation of the aquifer. It appears that no operational rules are applied to the wells.

In the fringe lands, the water table is shallow because the discharge capacity of the aquifer diminishes here for two reasons: (1) the hydraulic gradient reduces where the sloping alluvial fan reaches the flat desert area, and (2) the thickness and hydraulic conductivity of the aquifer diminishes. The necessary drainage canals for watertable control at the fringes of the irrigation perimeter are not maintained by the water authority, but by the respective farmers groups. For the irrigation water, these groups depend (1) on occasional river floods too large to be handled by the irrigation system and that flow down to the fringe lands through the natural watercourses, (2) on spillage from the irrigation system, and (3) on deep wells.

To stabilize the agriculture in the fringe lands, which are threatened by soil salinization, a method of strip-cropping can be recommended for soil salinity control. This method uses irrigated strips next to permanently un-irrigated strips, whereby the salinization is directed to the un-irrigated strips. This concept is sometimes called *sacrificial drainage*.

Punata

The alluvial fan of Punata is found in the district of Cochabamba, Bolivia. The region of Punata, at the upper end of the Valle (valley) Alto, at about 2800 m altitude, has a summer rainfall of 400 to 450 mm starting in the second half of November end ending in March. Maize is here the most important food crop, followed by potatoes. Alfalfa is the dominant fodder crop, followed by maize straw. These crops could, of old, only be planted successfully because of the existence of additional water resources like runoff, floods, river base-flow and groundwater. In the winter months, crop growth is restricted due to the occurrence of night frosts, especially in June and July, and absence of rains.

The total rural population in Punata is estimated at 25 000. There are about 4000 families of which an estimated 3680 are farmer families. The farms are small. The average size is 1.3 ha of which 1 ha is cropped. The modal size of farm is smaller, about 0.7 ha.

The rainfall distribution in Punata is characterized by a wet season from December to March, a dry season from May to October, and transition months in April and November. The average yearly total is 428 mm (1966 to 1983, San Benito). The rainfall with a probability of exceedance of 75% (R75) on a year basis is 360 mm. Rainfall is not reliable: in the period from 1966 to 1983, the yearly total varied between 246 mm (1982/83) to 591 mm (1968/69).

The river floods during the rainy summer period can be used for irrigation by anyone who wants to. When the river flow recedes, the stream can only be used for rotational irrigation by those who are entitled to take part in it (this is locally called the *mita* system). By the month of May the river base-flow becomes strongly reduced, and a drought period sets in, lasting into November.

Irrigation is considered desirable to start the cropping season in August/September, so that an early harvest can be obtained. The early harvest has a high market value and reduces peak labour requirements. Further, the irrigation reduces the risk of crop failure and it permits diversification of agricultural produce. Nevertheless, there are some farming communities that have refrained in the past from the extra effort to obtain additional irrigation water and who seemed to be content with purely rain-fed cropping. At a modest scale, irrigation from deep-wells is also practiced.

In order to satisfy the needs of the majority of the farmers who strongly wish to have additional irrigation water, the irrigation project Punata-Tiraque began to be developed from 1970 onwards. The project entailed the construction of a complicated system of dams and reservoirs up in the Andean mountains.

The gross area of the Punata projects is estimated at 4600 ha, 90% of which can be used for agriculture or animal husbandry. About 1150 ha of this presently receive irrigation water, either surface water derived from the *Laguna Robada* or *Lluska Kocha* dam, or water pumped from the 16 deep wells in the project area (estimated at 350 ha). In addition there are a few hundred hectares that receive occasional water from *mita* irrigation (wild flooding).

The traditional irrigation method is based on handling large irrigation flows (*golpes*) per farm at large intervals. The intake structures in the *Pucara Mayu* river, at the place where it enters the alluvial fan of Punata, would alternately pass water from each of the reservoir systems (*Laguna Robada* and *Lluska Kocha/Muyu Loma*) and the natural *mita* water. The new system has been designed for smaller flows with shorter rotation intervals, but it works continuously for the whole area, so that there is no need anymore to separate the various sources of water. It covers a much larger area than the traditional system and it incorporates the associations of the *mita* systems (which may have partly the same members), the associations of tube-well systems (which may also have partly the same members) as well as the persons who had no previous water rights.

Hence, the new irrigation system makes it necessary to replace the traditional water rights by a totally new set of rights (and duties). In addition, the farmers will have to get used to new water distribution methods and new field irrigation techniques. Because the new irrigation zones do not correspond to the boundaries of the existing, scattered, *Comité's de Riego*, not only the water management but also the organizational structure will have to be adjusted to the new situation.

Okavango

Features

The inland Okavango Delta in northwest Botswana is hand-shaped, with the fingers spread. The Okavango River, which originates in Angola, enters the delta at its apex. On the average, the river carries about 10 000 million m³ of water a year into the delta. The flow rate is high in the months of March and April (about 1000 m³/s on the average), but varying from year to year between 500 to 1500 m³/s and low in November (100 to 200 m³/s).

The large volume of water spreading over the delta is almost fully absorbed in permanent and seasonal swamps (the latter are called *molapo's*) before is slowly infiltrates or evaporates. There is rich swamp vegetation, which creates an ideal environment for numerous kinds of animals. The rich fauna finds its habitat on and between the thousands of islands between the swamps.

The little water that exceeds the retention capacity of the marshy wilderness drains from July to November through the fingers of the giant hand. Hence, it takes almost six months before the peak discharge of the Okavango River manifests itself at the base of the delta. Here, the water meets a barrier: the Thamalakane Fault Line, beyond which the Kalahari sands rise up 10 m. At the foot of the fault, the Thamalakane River collects the water (which is not more than 5% of the total inflow) and carries it almost without gradient to the Boteti River, which flows through a breach in the fault line. Eventually, the remaining waters evaporate in the Makgadikgadi Pans, more than 200 km to the east.

Although the annual rainfall is relatively low (an average of 500 mm, the greater part of which falls from December to March), it contributes a volume of water to the Delta equalling half of the Okavango inflow. The annual rainfall and its yearly distribution are equally erratic as the regime of the river.

The Okavango River transports a large amount of sands and other sediments into the delta. Their mass is about 2 million tons a year. Salts also enter the Delta, but they do so in a dissolved form. The salt concentration of the water is some 200 mg/l, which is very low. The total weight of incoming salts is thus about 2 million tons per year.

The sediments and the salts imported by the Okavango River settle in the delta. Together with the vegetation the sediments build up resistances to surface-water flow. As a result, the main watercourses have in the past swayed from thumb to little finger to and fro, as is common in alluvial fans. Tectonic movements have also contributed to this phenomenon. At present the middle finger, from which the Boro River stems, provides the major thoroughfare.

Many of the islands in the delta have a garland of riparian trees along their borders, but in the middle they are bare: symptoms of salt accumulation.

The Kalahari Desert cooperates with the Okavango River to form the predominantly sandy soils of the delta. The desert uses the vector wind to deposit its share of fine sand.

The geophysical characteristics of the Okavango Delta have led to a low population density, so that the natural situation has scarcely been disrupted by mankind. Also, the population was more interested in hunting and cattle breeding than in food crop production, so that agricultural developments were limited. The arable lands in the south-eastern fringes of the delta, that become dry after the floods recede (these are locally known as molapo's), often have a sandy topsoil. In depressions, the topsoil may be thin or missing altogether, exposing a heavy clay soil.

Developments

In 1978/79, after four years of high and prolonged floods had made *molapo* farming impossible, a severe drought coincided with an outbreak of foot and mouth disease, leaving the local population in a state of emergency. This resulted in two important undertakings:

Hydrographs of the flood level of the Boro river, with an indication of the years in which the timely closure of the sluice gates would facilitate flood-recession cropping in the molapo's;

- The first took place from 1979 to 1981 when, as part of a drought-relief program (Food for Work), FAO organized labour-intensive works to rehabilitate the flood-control bunds that the local population had built to protect their crops against inundations from rainstorms. Some new bunds were also built.

- The second was the construction of the "Buffalo-fence" which separates the outer fringes of the delta from its interior to prevent the spread of cattle diseases – especially foot and mouth disease. Completed in 1983, this fence has increased the importance of the *molapo's* outside it, and for the following reasons. Especially in the years when rains start late, the new grass flush in the *molapo's* after flood-recession presents about the only source of fodder in the region, also for the herds in the wide surroundings of the delta. With the *molapo's* inside the fence closed to cattle grazing, the grazing intensity in the *molapo's* outside the fence has increased.

The Molapo Development Projects (MDP) became operational in December 1983. The project aimed at increasing crop production in pilot areas by protecting those areas against prolonged high floods by flood-control bunds with entrance gates that could be closed. When enough floodwater has entered the molapo, the gate can be closed and the water recedes under influence of evaporation and infiltration into the soil and cropping may start when the outside water level is still high. This was a response to the high and prolonged flooding in the years 1974-1978, when *molapo* cropping was largely impossible. In more recent years, however, inadequate floods appeared to present an equally severe constraint to satisfactory crop production. It was therefore decided to focus also on improved, more stable crop production under fully rain-dependent conditions.

After the flood has been permitted to enter the bunded molapo, the sluice gates are closed. Recession of the water in the bunded molapo then begins under the influence of evaporation and infiltration, and allows a timely planting of the crop (in October or November). The crops use the residual soil moisture (about 100 mm) till the onset of the rainy season at the end of November or the beginning of December. Thus the growing season is prolonged, the moisture availability is increased, and crop production is enhanced. However, the success of the flood-control measures on the crop performance still depends to a great extent on the amount and distribution of the rainfall.

Drainage

Drainage is the natural or artificial removal of surface and subsurface water from an area. Many agricultural soils need drainage to improve production or to manage water supplies.

Early History

The earliest archaeological record of an advanced system of drainage comes from the Indus Valley Civilization from around 3100 BC in what is now Pakistan and North India. The ancient Indus systems of sewerage and drainage that were developed and used in cities throughout the civilization were far more advanced than any found in contemporary urban sites in the Middle East and even more efficient than those in some areas of modern Pakistan and India today. All houses in the major cities of Harappa and Mohenjodaro had access to water and drainage facilities. Waste water was directed to covered drains, which lined the major streets.

Drainage in the 19th Century

This operation is always best performed in spring or summer, when the ground is dry. Main drains ought to be made in every part of the field where a crosscut or open drain was formerly wanted; they ought to be cut four feet (1.2 m) deep, upon an average. This completely secures them from the possibility of being damaged by the treading of horses or cattle, and being so far below the small drains, clears the water finely out of them. In every situation, pipe-turfs for the main drains, if they can be had, are preferable. If good stiff clay, a single row of pipe-turf; if sandy, a double row. When pipe-turf cannot be got conveniently, a good wedge drain may answer well, when the subsoil is a strong, stiff clay; but if the subsoil be only moderately so, a thorn drain, with couples below, will do still better; and if the subsoil is very sandy, except pipes can be had, it is in vain to attempt under-draining the field by any other method. It may be necessary to mention here that the size of the main drains ought to be regulated according to the length and declivity of the run, and the quantity of water to be carried off by them. It is always safe, however, to have the main drains large, and plenty of them; for economy here seldom turns out well.

Having finished the main drains, proceed next to make a small drain in every furrow of the field if the ridges formerly have not been less than fifteen feet (5 m) wide. But if that should be the case, first level the ridges, and make the drains in the best direction, and at such a distance from each other as may be thought necessary. If the water rises well in the bottom of the drains, they ought to be cut three feet (1 m) deep, and in this ease would dry the field sufficiently well, although they were from twenty-five to thirty feet (8 to 10 m) asunder; but if the water does not draw well to the bottom of the drains, two feet (0.6 m) will be a sufficient deepness for the pipe-drain, and two and a half feet (1 m) for the wedge drain. In no case ought they to be shallower where the field has been previously levelled. In this instance, however, as the surface water is carried off chiefly by the water sinking immediately into the top of the drains, it will be necessary to have the drains much nearer each other—say from fifteen to twenty feet (5 to 6 m). If the ridges are more than fifteen feet (5 m) wide, however broad and irregular they may be, follow invariably the line of the old furrows, as the best direction for the drains; and, where they are high-gathered ridges, from twenty to twentyfour inches will be a sufficient depth for the pipe-drain, and from twentyfour to thirty inches for the wedge-drain. Particular care should be taken in connecting the small and main drains together, so that the water may have a gentle declivity, with free access into the main drains.

When the drains are finished, the ridges are cleaved down upon the drains by the plough; and where they had been very high formerly, a second clearing may be given; but it is better not to level the ridges too much, for by allowing them to retain a little of their former shape, the ground being lowest immediately where the drains are, the surface water collects upon the top of the drains; and, by shrinking into them, gets freely away. After the field is thus finished, run the new ridges across the small drains, making them about ten feet (3 m) broad, and continue afterwards to plough the field in the same manner as dry land. It is evident from the above method of draining that the expense will vary very much, according to the quantity of main drains necessary for the field, the distance of the small drains from each other, and the distance the turf is to be carried.

The advantage resulting from under-draining, is very great, for besides a considerable saving annually of water furrowing, cross cutting, etc., the land can often be ploughed and sown to advantage, both in the spring and in the fall of the year, when otherwise it would be found quite impracticable; every species of drilled crops, such as beans, potatoes, turnips, etc., can be cultivated successfully; and every species, both of green and white crops, is less apt to fail in wet and untoward seasons. Wherever a burst of water appears in any particular spot, the sure and certain way of getting quit of such an evil is to dig hollow drains to such a depth below the surface as is required by the fall or level that can be gained, and by the quantity of water expected to proceed from the burst or spring. Having ascertained the extent of water to be carried off, taken the necessary levels, and cleared a mouth or loading passage for the water, begin the drain at the extremity next to that leader, and go on with the work till the top of the spring is touched, which probably will accomplish the intended object. But if it should not be completely accomplished, run off from the main drain with such a number of branches as may be required to intercept the water, and in this way disappointment will hardly be experienced. Drains, to be substantially useful, should seldom be less than three feet (1 m) in depth, twenty or twenty four inches thereof to be close packed with stones or wood, according to circumstances. The former are the best materials, but in many places are not to be got in sufficient quantities; recourse therefore, must often be made to the latter, though not so effectual or durable.

It is of vast importance to fill up drains as fast as they are dug out; because, if left open for any length of time, the earth is not only apt to fall in but the sides get into a broken, irregular state, which cannot afterwards be completely rectified. A proper covering of straw or sod should be put upon the top of the materials, to keep the surface earth from mixing with them; and where wood is the material used for filling up, a double degree of attention is necessary, otherwise the proposed improvement may be effectually frustrated.

The pit method of draining is a very effectual one, if executed with judgment. When it is sufficiently ascertained where the bed of water is deposited, which can easily be done by boring with an auger, sink a pit into the place of a size which will allow a man freely to work within its bounds. Dig this pit of such a depth as to reach the bed of the water meant to be carried off; and when this depth is attained, which is easily discerned by the rising of the water, fill up the pit with great land-stones and carry off the water by a stout drain to some adjoining ditch or mouth, whence it may proceed to the nearest river.

Current Practices

Modern drainage systems incorporate geotextile filters that retain and prevent fine grains of soil from passing into and clogging the drain. Geotextiles are synthetic textile fabrics specially manufactured for civil and environmental engineering applications. Geotextiles are designed to retain fine soil particles while allowing water to pass through. In a typical drainage system they would be laid along a trench which would then be filled with coarse granular material: gravel, sea shells, stone or rock. The geotextile is then folded over the top of the stone and the trench is then covered by soil. Groundwater seeps through the geotextile and flows within the stone to an outfall. In high groundwater conditions a perforated plastic (PVC or PE) pipe is laid along the base of the drain to increase the volume of water transported in the drain.

Alternatively, prefabricated plastic drainage systems made of HDPE called Smart Ditch, often incorporating geotextile, coco fiber or rag filters can be considered. The use of these materials has become increasingly more common due to their ease of use which eliminates the need for transporting and laying stone drainage aggregate which is invariably more expensive than a synthetic drain and concrete liners. Over the past 30 years geotextile and PVC filters have become the most commonly used soil filter media. They are cheap to produce and easy to lay, with factory controlled properties that ensure long term filtration performance even in fine silty soil conditions.

21st Century Alternatives

Seattle's Public Utilities created a pilot program called Street Edge Alternatives (SEA Streets) Project. The project focuses on designing a system "to provide drainage that more closely mimics the natural landscape prior to development than traditional piped systems". The streets are characterized by ditches along the side of the roadway, with plantings designed throughout the area. An emphasis on non curbed sidewalks allows water to flow more freely into the areas of permeable surface on the side of the streets. Because of the plantings the run off water from the urban area does not all directly go into the ground but can also be absorbed into the surrounding environment. According to the monitoring by Seattle Public Utilities, they report at 99 percent reduction of storm water leaving the drainage project

Drainage in Construction

The civil engineer or site engineer is responsible for drainage in construction projects. They set out from the plans all the roads, Street gutters, drainage, culverts and sewers involved in construction operations. During the construction of the work on site he/she will set out all the necessary levels for each of the previously mentioned factors. Site engineers work alongside architects and construction managers, supervisors, planners, quantity surveyors, the general workforce, as well as subcontractors. Typically, most jurisdictions have some body of drainage law to govern to what degree a landowner can alter the drainage from his parcel.

Reasons for Artificial Drainage

Wetland soils may need drainage to be used for agriculture. In the northern USA and Europe, glaciation created numerous small lakes which gradually filled with humus to make marshes. Some of these were drained using open ditches and trenches to make mucklands, which are primarily used for high value crops such as vegetables.

The largest project of this type in the world has been in process for centuries in the Netherlands. The area between Amsterdam, Haarlem and Leiden was, in prehistoric times swampland and small lakes. Turf cutting (Peat mining), subsidence and shoreline erosion gradually caused the formation of one large lake, the Haarlemmermeer, or lake of Haarlem. The invention of wind powered pumping engines in the 15th century permitted drainage of some of the marginal land, but the final drainage of the lake had to await the design of large, steam powered pumps and agreements between regional authorities. The elimination of the lake occurred between 1849 and 1852, creating thousands of km² of new land.

Coastal plains and river deltas may have seasonally or permanently high water tables and must have drainage improvements if they are to be used for agriculture. An example is the flatwoods citrus-growing region of Florida. After periods of high rainfall, drainage pumps are employed to prevent damage to the citrus groves from overly wet soils. Rice production requires complete control of water, as fields need to be flooded or drained at different stages of the crop cycle. The Netherlands has also led the way in this type of drainage, not only to drain lowland along the shore, but actually pushing back the sea until the original nation has been greatly enlarged.

In moist climates, soils may be adequate for cropping with the exception that they become waterlogged for brief periods each year, from snow melt or from heavy rains. Soils that are predominantly clay will pass water very slowly downward, meanwhile plant roots suffocate because the excessive water around the roots eliminates air movement through the soil.

Other soils may have an impervious layer of mineralized soil, called a hardpan or relatively impervious rock layers may underlie shallow soils. Drainage is especially important in tree fruit production. Soils that are otherwise excellent may be waterlogged for a week of the year, which is sufficient to kill fruit trees and cost the productivity of the land until replacements can be established. In each of these cases appropriate drainage carries off temporary flushes of water to prevent damage to annual or perennial crops.

Drier areas are often farmed by irrigation, and one would not consider drainage necessary. However, irrigation water always contains minerals and salts, which can be concentrated to toxic levels by evapotranspiration. Irrigated land may need periodic flushes with excessive irrigation water and drainage to control soil salinity.

Tile Drainage

Tile drainage (in agriculture) is a practice for removing excess water from the subsurface of soil intended for agriculture. Whereas irrigation is the practice of adding additional water when the soil is naturally too dry, drainage brings soil moisture levels down for optimal crop growth. While surface water can be drained via pumping and/ or open ditches, tile drainage is often the best recourse for subsurface water. Too much subsurface water can be counterproductive to agriculture by preventing root development, and inhibiting the growth of crops. Too much water also can limit access to the land, particularly by farm machinery. In terms of access, most modern agriculture depends on the usage of large machinery—tractors and implements— to prepare the seedbed, plant the crop, carry out any cultivation and applications during the growing season, and ultimately, to harvest the crop. Operating most machinery in excessively wet conditions may result in soil degradation because of excessive soil compaction, and inhibit the operation of the machinery (i.e., "getting stuck").

Increased Crop Yields

Most crops require specific soil moisture conditions, and do not grow well in wet, "mucky" soil. Even in soil that isn't "mucky" the roots of most plants will not grow much deeper than the water table. Early in the growing season when water is in ample supply, plants are small and don't require much water. During this time, the plants do not need to develop their roots to "reach" for the water. As the plants grow and use more water during the growing season water becomes more scarce. During this time, the water table starts to fall. Plants suddenly need to start developing roots to reach to the water. During dry times, the water table can fall faster than the plants can develop roots to "reach" for the water. This can seriously stress the plants.

By adding drain tile, the water table is effectively lowered, and plants can properly develop their roots. The lack of water saturation allows oxygen to exist in the soil around the roots. Drain tile prevents the roots from being under the water table during wet periods that could cause excessive plant stress. By removing excessive water, crops use water they have more effectively.

An increase in crop yield can be simply summarized by the following: Simply by forcing the plants to have more developed roots, the plants can more effectively absorb more nutrients and water.

The same principal is employed by containers that hold house plants: they have drain holes in the bottom to allow oxygen to the roots, and prevent soil saturation. By placing drain tile under a field in a grid style layout, the same effect can be employed on a several hundred acre field.

Plumbing of Drain Tile

In a tile drainage system, a sort of "plumbing" is installed below the surface of agricultural fields, effectively consisting of a network of below-ground pipes that allow subsurface water to move out from between soil particles and into the tile line. Water flowing through tile lines is often ultimately deposited into surface water points—lakes, streams, and rivers—located at a lower elevation than the source. Water enters the tile line either via the gaps between tile sections, in the case of older tile designs, or through small perforations in modern plastic tile.

Soil type greatly affects the efficacy of tile systems, and dictates the extent to which the area must be tiled in order to ensure sufficient drainage. Sandier soils will need little, if any, additional drainage, whereas soils with high clay contents will hold their water tighter, requiring tile lines to be placed closer together.

History of Tile Drainage

Both Cato and Pliny have described tile drainage systems, in 200 BC and the first century AD, respectively. According to the Johnston Farm website, tile drainage was first introduced to the United States in 1838, when John Johnston brought the practice from his native Scotland to his farm in Seneca County, New York. Johnston laboured to lay 72 miles (116 km) worth of clay tile on 320 acres (1.3 km²). The effort paid off by increasing his wheat yield from 12 bushels per acre to 60 bu/acre. Johnston, "Father of American Tile Drainage", continued to advocate tile drainage throughout his life, attributing his success as a farmer to the formula "D,C, and D" (dung, credit, and drainage).

The expansion of drainage networks was an important technical aspect of Westward Expansion in the 19th century. Although land in the United States was parcelled out in accordance with the Public Land Survey System as established by the Land Ordinance of 1785, development, especially of agricultural land, was often limited by the rate at which it was made capable for cultivation. For example, although Iowa was made a state in 1846, maps depicting land ownership show below-average population densities in the northwestern region as late as the 1870s, a corner of the state which today is still noted for its high water table and numerous lakes, marshes, and wetlands.

States throughout the region faced similar limits to agricultural intensification. Many states offered government incentives to improve land for farming. For example, legislation in Indiana prompted an Act of Congress in 1850 that provided for swamplands to be sold at a discount to farmers on the condition that they drain the land and bring it into agricultural productivity. To facilitate this process, most states set up government agencies to oversee and regulate the installation of tile drainage systems. Even today, ballots for elections in rural America often include candidates for local drainage supervisory boards. The decades following the American Civil War saw rapid expansion of drainage systems. For example, historical literature from Ohio notes that in the year of 1882, the number of acres drained was about equal to the area of land drained in all previous years. In the 1930s, the Civilian Conservation Corps contributed to the tile network throughout the Midwest, much of which is still in use.

Advances in Drainage Technology

Throughout the twentieth century, the technology of tile installation remained similar to the methods first used in 1838. Although cement sections later replaced the original clay tiles, and machines were used to dig the trenches for the tile lines, the process remained quite labour-intensive and limited to specialized contractors.

The introduction of plastic tile served to reduce both the cost of tile installation, as well as the amount of labour involved. Rather than set individual sections of cement tile end-to-end in the trench, tile installers had only to unroll a continuous section of lightweight, flexible tile line. Towards the end of the twentieth century, when large, four-wheel-drive tractors became more common on American farms, do-it-yourself tile implements appeared on the market. By making tile installation cheaper and allowing it to be done on the landowner's schedule, farmers are capable of draining localized wet spots that may not create enough of a problem to merit more costly operations. In this way, farmers may enjoy increases in crop yield while saving on the capital costs of tile installation.

Social and Ecological Effects of Tile Drainage

Unfortunately, the ability for farmers to install their own tile can be problematic. First of all, private installations may reduce the ability of local drainage supervisory boards to regulate tile installation, which in some areas of the country requires proper documentation before a contractor can continue. This leads to the second potential conflict, the unintentional interruption of existing tile networks. Most "do-it-yourself" tile flows do not dig trenches but rather split the soil enough to squeeze the tile line in; thus, a farmer would not be aware if he breaks a line of tile that might serve his neighbors, as well (Mutual tile lines are often dictated by topography rather than land ownership, and the location of many old, but still effective, tile lines are unknown.). The potential for across-the-fence disputes are obvious.

Ecologically, the expansion of drainage systems has had tremendous negative effects. Hundreds of thousands of wetland species experienced significant population declines as their habitat was increasingly fragmented and destroyed. Although market hunting within the Central Flyway was a contributing factor in the decline of many waterfowl species' numbers in the early decades of the twentieth century, loss of breeding habitat to agricultural expansion is certainly the most significant. Early maps of midwestern states depict many lakes and marshes that are either nonexistent or significantly reduced in area today. Channelization, a related process of concentrating and faciliting the flow of water from agricultural areas, also contributed to this degradation.

In bypassing the natural flow of water from the surface to the water table, drainage systems often prevent the natural filtration of water provided by soils and wetlands. Thus, drainage systems pose a threat to surface water sources by directly depositing water laden with fertilizers, eroded soil, agrochemicals, and other types of agricultural runoff pollutants. In very flat areas, where the natural topography does not provide the gradient necessary for water flow, "agricultural wells" can be dug to provide tile lines sufficient outlet. In these cases, it is the groundwater that stands to be polluted by unfiltered tile output.

Intensive Livestock Operations (ILO) have led to challenges of livestock effluent disposal. Livestock effluent contains valuabe nutrients, but the misapplication of these materials can lead to serious ecological problems, such as nutrient loading. Injecting effluent directly into the ground is one method employed by manure applicators to improve nutrient uptake. Drainage tiles may increase injected manure seepage into surface waterways from manure injection because liquid manure seeps through soils and then drains out of the field and into waterways via drainage tiles.

Today, a number of state and federal initiatives serve to reverse habitat loss. Many programs encourage and even reimburse farmers for interrupting the drainage of localized wetholes on their property, often by breaking tile intakes or removing the tile completely. Landowners are often partially or fully compensated for forfeiting the ability to grow crops on this land. Such programs and the cooperation of landowners across the country have had significant positive effects on the populations of a wide variety of waterfowl.

Drainage System (Agriculture)

An agricultural drainage system is a system by which the water level on or in the soil is controlled to enhance agricultural crop production.

Classification

The function of the field drainage system is to control the water table, whereas the function of the main drainage system is to collect, transport, and dispose of the water through an outfall or outlet. In some instances one makes an additional distinction between collector and main drainage systems.

Field drainage systems are differentiated in surface and subsurface field drainage systems. Sometimes (e.g. in irrigated, submerged rice fields), a form of temporary drainage is required whereby the drainage system is allowed to function on certain occasions only (e.g. during the harvest period). If allowed to function continuously, excessive quantities of water would be lost. Such a system is therefore called a checked, or controlled, drainage system. More usually, however, the drainage system is meant to function as regularly as possible to prevent undue waterlogging at any time and one employs a regular drainage system. In literature, this is sometimes also called a "relief drainage system".

Surface Drainage Systems

The regular surface drainage systems, which start functioning as soon as there is an excess of rainfall or irrigation, operate entirely by gravity. They consist of reshaped or reformed land surfaces and can be divided into:

 • Bedded systems, used in flat lands for crops other than rice;
 • Graded systems, used in sloping land for crops other than rice.

The bedded and graded systems may have ridges and furrows.

The checked surface drainage systems consist of check gates placed in the embankments surrounding flat basins, such as those used for rice fields in flat lands. These fields are usually submerged and only need to be drained on certain occasions (e.g. at harvest time). Checked surface drainage systems are also found in terraced lands used for rice.

In literature, not much information can be found on the relations between the various regular surface field drainage systems, the reduction in the degree of waterlogging, and the agricultural or environmental effects. It is therefore difficult to develop sound agricultural criteria for the regular surface field drainage systems. Most of the known criteria for these systems concern the efficiency of the techniques of land levelling and earthmoving.

Similarly, agricultural criteria for checked surface drainage systems are not very well known.

Subsurface Drainage Systems

Like the surface field drainage systems, the subsurface field drainage systems can also be differentiated in regular systems and checked (controlled) systems.

When the drain discharge takes place entirely by gravity, both types of subsurface systems have much in common, except that the *checked* systems have control gates that can be opened and closed according to need. They can save much irrigation water. A *checked* drainage system also reduces the discharge through the main drainage system, thereby reducing construction costs.

When the discharge takes place by *pumping*, the drainage can be checked simply by not operating the pumps or by reducing the pumping time. In northwestern India, this practice has increased the irrigation efficiency and reduced the quantity of irrigation water needed, and has not led to any undue salinization.

The subsurface field drainage systems consist of horizontal or slightly sloping channels made in the soil; they can be open ditches, trenches, filled with brushwood and a soil cap, filled with stones and a soil cap, buried pipe drains, tile drains, or mole drains, but they can also consist of a series of wells.

Modern buried pipe drains often consist of corrugated, flexible, and perforated plastic (PE or PVC) pipe lines wrapped with an *envelope* or filter material to improve the permeability around the pipes and to prevent entry of soil particles, which is especially important in fine sandy and silty soils. The surround may consist of synthetic fibre (geotextile).

The *field drains* (or *laterals*) discharge their water into the collector or main system either by *gravity* or by *pumping*.

The wells (which may be open dug wells or *tubewells*) have normally to be pumped, but sometimes they are connected to drains for discharge by gravity. Subsurface drainage by wells is often referred to as vertical drainage, and drainage by channels as horizontal drainage, but it is more clear to speak of "field drainage by wells" and "field drainage by ditches or pipes" respectively.

In some instances, subsurface drainage can be achieved simply by breaking up slowly permeable soil layers by *deep plowing* (*subsoiling*), provided that the underground has sufficient natural drainage. In other instances, a combination of sub-soiling and subsurface drains may solve the problem.

Main Drainage Systems

The main drainage systems consist of *deep* or *shallow* collectors, and *main drains* or *disposal drains*.

Deep collector drains are required for subsurface field drainage systems, whereas shallow collector drains are used for surface field drainage systems, but they can also be used for pumped subsurface systems. The deep collectors may consist of open ditches or buried pipe lines. The terms *deep collectors* and *shallow collectors* refer rather to the depth of the water level in the collector below the soil surface than to the depth of the bottom of the collector. The bottom depth is determined both by the depth of the water level and by the required discharge capacity.

The deep collectors may either discharge their water into deep main drains (which are drains that do not receive water directly from field drains, but only from collectors), or their water may be pumped into a disposal drain.

Disposal drains are main drains in which the depth of the water level below the soil surface is not bound to a minimum, and the water level may even be above the soil surface, provided that embankments are made to prevent inundation. Disposal drains can serve both subsurface and surface field drainage systems. Deep main drains can gradually become disposal drains if they are given a smaller gradient than the land slope along the drain. The technical criteria applicable to main drainage systems depend on the hydrological situation and on the type of system.

Main Drainage Outlet

The final point of a main drainage system is the gravity outlet structure or the pumping station.

Applications

Surface drainage systems are usually applied in relatively flat lands that have soils with a low or medium infiltration capacity, or in lands with high-intensity rainfalls that exceed the normal infiltration capacity, so that frequent waterlogging occurs on the soil surface.

Subsurface drainage systems are used when the drainage problem is mainly that of shallow water tables.

When both surface and subsurface waterlogging occur, a combined surface/subsurface drainage system is required.

Sometimes, a subsurface drainage system is installed in soils with a low infiltration capacity, where a surface drainage problem may improve the soil structure and the infiltration capacity so greatly that a surface drainage system is no longer required.

On the other hand, it can also happen that a surface drainage system diminishes the recharge of the groundwater to such an extent that the subsurface drainage problem is considerably reduced or even eliminated.

The choice between a subsurface drainage system by pipes and ditches or by tube wells is more a matter of technical criteria and costs than of agricultural criteria, because both types of systems can be designed to meet the same agricultural criteria and achieve the same benefits. Usually, pipe drains or ditches are preferable to wells. However, when the soil consists of a poorly permeable top layer several meters thick, overlying a rapidly permeable and deep subsoil, wells may be a better option, because the drain spacing required for pipes or ditches would be considerably smaller than the spacing for wells.

When the land needs a subsurface drainage system, but saline groundwater is present at great depth, it is better to employ a shallow, closely spaced system of pipes or ditches instead of a deep, widely spaced system. The reason is that the deeper systems produce a more salty effluent than the shallow systems. Environmental criteria may then prohibit the use of the deeper systems.

In some drainage projects, one may find that only main drainage systems are envisaged. The agricultural land is then still likely to suffer from field drainage problems. In other cases, one may find that field drainage systems are ineffective because there is no adequate main drainage system. In either case, the installation of drainage systems is not recommended.

Reference: Gives a general description of land drainage in the world and shows a paper on types of agricultural land drainage systems used in different parts of the world.

Drainage System Design

The analysis of positive and negative (side) effects of drainage and the optimization of drainage design in accordance to the *drainage design procedures* is discussed in the article on Drainage research.

Watertable Control

Watertable control is the practice of controlling the water table in agricultural land by subsurface drainage with proper criteria to improve the crop production.

Description and Definitions

Subsurface land drainage aims at controlling the water table of the groundwater in originally waterlogged land at a depth acceptable for the purpose for which the land is used. The depth of the water table with drainage is *greater* than without.

Purpose

In agricultural land drainage, the purpose of water table control is to establish a depth of the water table that does no longer interfere negatively with the necessary farm operations and crop yields. In addition, land drainage can help with soil salinity control. The soil's hydraulic conductivity plays an important role in drainage design.

Optimization

Optimization of the depth of the water table is related the benefits and costs of the drainage system. The shallower the permissible depth of the water, the lower the cost of the drainage system to be installed to achieve this depth. However, the lowering of the originally too shallow depth by land drainage entails *side effects*. These have also to be taken into account, including the costs of mitigation of negative side effects.

The *optimization* of drainage design and the development of *drainage criteria* are discussed in the article on drainage research.

Example of the effect of drain depth on soil salinity and various irrigation/drainage parameters as simulated by the Salt Mod program.

History

Historically, agricultural land drainage started with the digging of relatively shallow open ditches that received both runoff from the land surface and outflow of groundwater. Hence the ditches had a surface as well as a subsurface drainage function.

By the end of the 19th century and early in the 20th century it was felt that the ditches were a hindrance for the farm operations and the ditches were replaced by buried lines of clay pipes (tiles), each tile about 30 cm long. Hence the term "tile drainage".

Since 1960, one started using long, flexible, corrugated plastic (PVC or PE) pipes that could be installed efficiently in one go by trenching machines. The pipes could be pre-wrapped with an envelope material, like synthetic fibre and geotextile, that would prevent the entry of soil particles into the drains.

Thus, land drainage became a powerful industry. At the same time agriculture was steering towards maximum productivity, so that the installation of drainage systems came in full swing.

Environment

As a result of large scale developments, many modern drainage projects were *over-designed*, while the negative environmental side effects were ignored. In circles with environmental concern, the profession of land drainage got a poor reputation, sometimes justly so, sometimes unjustified, notably when land drainage was confused with the more encompassing activity of wetland reclamation. Nowadays, in some countries, the hardliner trend is reversed. Further, *checked* or *controlled* drainage systems were introduced, as shown in discussed on the page: Drainage system (agriculture).

Drainage Design

The design of subsurface drainage systems in terms layout, depth and spacing of the drains is often done using subsurface drainage equations with parameters like drain depth, depth of the water, soil depth, hydraulic conductivity of the soil and drain discharge. The drain discharge is found from an agricultural water balance.

Drainage by Wells

Subsurface drainage of groundwater can also be accomplished by pumped wells (*vertical* drainage, in contrast to *horizontal* drainage). Drainage wells have been used extensively in the Salinity Control and Reclamation Program (SCARP) in the Indus valley of Pakistan.

Although the experiences were not overly successful, the feasibility of this technique in areas with deep and permeable aquifers is not to be discarded. The well spacings in these areas can be so wide (more than 1000m) that the installation of *vertical* drainage systems could be relatively cheap compared to *horizontal* subsurface drainage (drainage by pipes, ditches, trenches, at a spacing of 100m or less). For the design of a well field for control of the water, the Well Drain model may be helpful.

Participatory Irrigation Management

A Process of Improving Productivity and Sustainability of Irrigation Systems

Participatory Irrigation Management (PIM) refers to the involvement of irrigation users in all aspects and all levels of irrigation management.

"All aspects" includes the initial planning and design of new irrigation projects or improvements, as well as the construction, supervision, and financing, decision rules, operation, maintenance, monitoring, and evaluation of the system.

What are the Problems of Irrigation

In a national seminar on PIM in India held in 1994, the participants identified the following problems, or issues, with irrigation management. While one can debate whether some of these are "problems" or "symptoms" of problems, they are certainly issues familiar to all of us:

- Inadequate water availability at the lowest outlets
- Poor condition/maintenance of the system
- Lack of measuring devices and control structures
- Inadequate allocation for O&M
- Inequitable distribution of water
- Lack of incentives for saving water
- Poor drainage.

In particular countries and particular irrigation systems, some problems may be more critical than others. There are always additional problems or issues that could be added.

How can Participation Improve Irrigation?

Now let us consider the problems with irrigation management and think about whether and how participatory management approaches can help solve them. We will consider them one by one.

- Inadequate water availability at the lowest outlets
- ? Poor condition/maintenance of the system
- Lack of measuring devices and control structures
- Inadequate allocation for O&M
- Inequitable distribution of water
- Lack of incentives for saving water
- Poor drainage.

This new perspective of irrigation management has been applied to both projects as well as policies and has greatly expanded our understanding of why some irrigation systems work better than others. It is not only a physical problem; irrigation also involves a much broader set of management issues. For example, how is the system financed? What type of monitoring is carried out and how is this fed back into the O&M process?

Indeed, if we do a straight line projection of what is happening, we would perhaps predict that the state will eventually disappear entirely from the irrigation over the next several decades. But no one is realistically predicting the demise of the state's role in managing irrigation; there will continue to be an essential management role for the state. What is happening is a rationalization of the respective roles of government and users. The pendulum that had shifted far towards the direction of strong government involvement in all aspects of irrigation management is now swinging the other way, towards a more sensible equilibrium.

Inadequate Water Availability at the Lowest Outlets

If the water availability is inadequate because there is inadequate supply, then there are clear limits within which to suggest improvements. But even here, participation in the sense of transparency and dissemination of information about the water situation (e.g., through a system-wide organization) can engender good will from farmers vis a vis the system managers, thus minimizing the likelihood of intentionally breaking gates to increase water supply.

When the users understand the constraints faced by management, they are far more likely to cooperate. If the water availability at the lowest outlets is inadequate because of water thefts upstream, then part of the solution could entail bringing upstream and downstream users into a single management entity that would be accountable to all the users. This principle of including as many stakeholders as possible in a single organization underlies the very large water user organizations found in Mexico, for example. Participation of the users in managing the system is not the only possible solution, of course. More diligent management by government irrigation engineers could also solve the problem. But what are the incentives to an government engineer who will be paid by his agency at the same (usually low) rate regardless of how diligently he supervises water use between the upstream and downstream users.

Compare this situation with that of an irrigation engineer hired by a water user organization to oversee water distribution between upstream and downstream users. If he does not satisfy both sets of farmers, he will lose his job, and, at least in the case of Mexico, his salary may be double that of his agency counterpart. This engineer has a very strong incentive to distribute water equitably throughout the system, so that he remains in good favour in the eyes of all farmers.

Poor Condition/Maintenance of the System

The almost universal problem of deferred maintenance resulting in deteriorating infrastructure is on everyone's list of irrigation management concerns. Can participation help? When the users of the system are also in charge of maintenance, they have every incentive to perform timely repairs, to monitor the quality of the work, and to protect structures from vandalism. When the users also participate in initial design decisions, with the understanding that they will inherit the system as their own, they have a strong incentive to demand a system whose operation poses no technical problems, to monitor construction quality, and if they will repay some of the capital costs, to minimize the overall costs of construction. Most often this translates into reduced demand for lining canals except where it makes good operational sense.

The conservative tendency on lining which participation encourages may in itself result in substantial cost savings initially, and long-term improvements in maintenance (since users are reluctant to repair lining that they do not consider essential).

Lack of Measuring Devices and Control Structures

When irrigation managers are engineers employed by the government irrigation agency, there is little incentive to measure actual water flows. The philosophy of "no news is good news" is the standard approach to measurement. Similarly, control structures require both operational attention and maintenance which may be beyond the interest and/or budgets of the managing engineer. Would this situation change if farmers were in charge of management?

Imagine a situation in which the managing engineer would have to prepare weekly reports to a governing board of farmers, to whom the engineer owes his job. Would he now have an incentive to know how much water is going into each outlet? Would he now have an incentive to control the water so that he can adhere to his operational plan? This type of performance-based management by which the governing board holds the engineer accountable for water flows is theoretically possible under a government administered system. But if we can imagine two governing boards of an irrigation system, the first comprised of elected farmers who themselves depend of that irrigation water, and the second comprised of civil servants who enjoy job security, which of these boards is more likely to demand high performance? Which of these boards would be more likely to be interested in the flow rates in various reaches of the system?

Inadequate Allocation for O&M

Inadequate allocation implies either high costs, or low recovery, or both. The costs of both operating and maintaing the system are increased when farmers deliberately intervene to disrupt the scheduled operation, or to damage structures, and costs are reduced when farmers respect the operating plan and report maintenance problems promptly. Cost recovery appears to be enhanced when farmers can see that the money is going into system O&M (see Vicious Cycle of Public Irrigation), and when they perceive that they are receiving valuable services from the system, i.e., the timely provision of water. Can participatory irrigation management contribute to lower costs and higher recovery? When farmers are directly involved in the design of the system itself and the placement of structures and when they have input into establishing the operational plan, they will be more likely to both understand and protect the facilities and the plan.

When farmers are clearly the owners of the physical system, so that the maintenance costs are their own responsibility, they will have a strong incentive to protect the physical integrity of the system to reduce their overall costs. The cost recovery issue is complex. When farmers are involved in management they have a better understanding of how the funds are being used, and this knowledge is likely to render them more willing to pay the fees. To the extent that irrigation management is improved with user involvement, the farmers may feel more willing to pay for improved service (although these links are difficult to prove). Perhaps the best evidence for a relationship between participatory management and cost recovery comes from actual experience. In the Philippines, the payment rate for irrigation service fees is significantly higher in those systems where water user associations have been formed. In Mexico, payment rates are close to 100% because the system managers — controlled by a board of irrigation users — will refuse to give water without prior payment. Political realities would not allow this type of ultimatum were the managers part of the Irrigation Department.

Inequitable Distribution of Water

Disparities in the supply of water available to farmers are not only due to the physical problems of reaching the tail end of large systems. Disparities may also be caused by intentional tampering with the system, either by physically enlarging outlets, or offering bribes to the water masters. The result is that influential farmers can take more than their due share, thus depriving others who have the misfortune of being located downstream. Can management participation by the water users help solve this well-known situation? Or would the withdrawal of state authority actually encourage powerful farmers to take even more water, with no fear of government sanctions? Social theory suggests that under user management, tampering with water distribution becomes much more difficult. There is greater transparency and accountability rendering illicit water deliveries more difficult to conceal and even more difficult to defend. Whereas thefts of "government" water might be tolerated by other farmers, when the water supply is allocated to the collectivity of farmers, any theft of water implies stealing from a fellow farmer, and a fellow member of the association.

Lack of Incentives for Saving Water

What are the incentives that can encourage farmers to conserve water? Can such incentives become operative more readily under user management or under government management? The set of factors that determine water conservation behaviour are complex. One factor is security of supply. If farmers know when they expect to irrigate next time, they may be more modest in applying water, whereas if they are afraid they might have to wait a long time for the next irrigation, they will be apt to over-irrigated as insurance against this uncertainty. Another factor may be information and management confidence. If farmers are aware that there is an overall scarcity of supply and they believe the scarce water will be equitably distributed, they will may be more likely to limit their own consumption, or at least refrain from taking more than their share. A third factor is the transferability of water rights. If farmers in one part of the command area do not need water this week, but may need extra water after three weeks, can they forgo water now and use it three weeks from now?

While these three types of incentives — security of supply, information and management confidence, and transferability of water rights — can be found under both government management and user management, they are more likely to be found under user management. The reasons have to do with accountability. When the manager are accountable to the users, those managers have a direct incentive to keep supplies regular, and to inform the users of any problems. The continued employment of the managers depends upon the management confidence of the users. Similarly, the transfer of water rights from one user to another, or from one association to another, is not impossible under government management, but it is unusual and perhaps non-existent. Water markets are the formal manifestation of transferring water rights. While water markets are very much a novelty in the irrigation sector, there is a promising potential for tapping market forces to provide economic incentives for saving water, and for using it in the most productive manner.

Poor Drainage

Investments in drainage are notoriously under-valued in government managed irrigation systems. The emphasis is nearly always on increasing the supply of irrigation water, rather than on facilitating disposal of drainage water. It would be unrealistic to claim that farmers are necessarily more foresighted than government agencies in giving attention to drainage, but there is evidence from Mexico that more attention is given to drainage after transfer to user management. Part of the reason in the Mexican case is that the user-managers are able to operate the irrigation system at an overall lower cost, and therefore have funds available for drainage O&M.

Another part of the reason is that the government agency continues to subsidize drainage through monitoring soil and water quality and helping the new managers address priority problems strategically. In Mexico the water user organizations are responsible for both irrigation and drainage. In the Chambal project in Rajasthan, India, separate drainage committees are being established to operate and maintain new investments (by the government) in on-farm drains.

Rationale for Participation

Why participation? Another question might also be asked: "Why should the government be involved in irrigation?" Clearly, there are investments that only the government can make, or where the government has a definite advantage vis-a-vis farmers, even very well organized associations of farmers.

Construction of dams and barrages, for example, or large canals, would be extremely difficult for farmers to handle. Governments provide us with available institutional resources — departments, agencies, trained staff, etc. — which can be used to get things done. Why re-invent the wheel and ask farmers to organize their own arrangements for building as dam?

 • Comparative advantages
 • Improved design, construction, and O&M.
 • Lower costs to government.
 • Social capital.

Comparative Advantage

Farmers have some comparative advantages as well. They have direct incentives to manage irrigation water in a productive and sustainable manner; they offer an on-the-ground presence that even the most dedicated off-site agency staff cannot equal, and they have an intimate knowledge about their fellow irrigators. The logic of the PIM approach is that both governments and farmers have separate comparative advantages. At the moment, governments are trying to do much more than they can do well. What are the advantages that management by farmers — by the users — can offer?

Improved Design, Construction and O&M

When farmers are directly involved in the design process, whether for new systems or rehabilitation of old ones, they will provide useful design input and they will come away with an understanding of the design logic of the system they will be managing.

During construction, farmer input has the functions of quality control (ensuring design standards are met), cost savings (through guarding against needless spending, and substituting some costs with farmers' own labour), and construction knowledge. Knowing how the system is constructed will help in repairs later on. The advantage of farmer inputs into O&M, either as direct managers or as the overseers of technical managers, has been discussed.

Lower Costs to Government

Cost savings to the government irrigation agency is often the driving force behind irrigation policy reforms. Government run systems are chronically short of maintenance funds leading to deteriorating systems and more difficult operation. Management transfer of major levels of the system to users offers government agencies an escape from this vicious cycle.

While some critics see this as merely passing the costs on to farmers, the picture is not usually so bleak. Evidence from Mexico and Turkey suggest that farmers can manage better and more cheaply than their government predecessors. Thus, both farmers and government can benefit from these cost savings; farmers can enjoy better service, and cost savings; the government incurs less management cost and can then afford to improve service in the main system.

Social Capital

The organizations that farmers establish for managing their irrigation systems constitute a form of social capital that can have spin-off effects in other aspects of social and economic life. The network of contacts among agency staff and the water user organization leadership, for example, can bring the farming community into closer touch with related services, e.g., credit, educations opportunities, or even political access. And the skills that farmers learn through their experience with their water user organization — accounting, budgeting, planning, organizing — constitute a set of knowledge that can be used in many other productive endeavours.

What is PIM?

- PIM means "Participation" — not only in O&M and financing, but in making decisions that will affect the O&M — and financing
- Continuum of management approaches and levels of participation
- The concept of users' organization and transfer of management to users is different from PIM.

PIM Means Participation

Participation refers to a continuum of involvement in management decisions. One meaning of "PIM" may be that the irrigation users have total control and responsibility over the operations and maintenance of part or all of the irrigation system. Another meaning of PIM may be that a farmer council plays an advisory role, with real power remaining in the hands of the irrigation agency. Various levels of participation are outlined below.

1. Information sharing
   - Translation into local languages and dissemination of written material using various media
   - Informational presentations and public meetings.

2. Consultations
   - Meetings
   - Field visits and interviews.

3. Joint Assessments
   - Participatory assessments and evaluations
   - Beneficiary assessments.

4. Shared decisionmaking
   - Participatory planning
   - Workshops and seminars to determine positions, priorities, roles
   - Meetings to resolve conflicts, seek agreements, engender ownership
   - Public reviews of draft documents.

5. Collaboration
   - Formation of joint agency/stakeholder committees/task forces
   - Joint work with user groups, NGOs, or other stakeholder groups
   - Stakeholder groups given principal responsibility for implementation.

6. Empowerment
   - Capacity building of stakeholder organizations
   - Hand-over and self-management by stakeholders
   - Support for new, spontaneous initiatives by stakeholders.

Continuum of Involvement in Management Decisions

We can characterize the range of state-user relationships as a continuum from the state doing everything on behalf of the users, to the case of the state doing nothing for the users, other than leaving them alone. In between these two ends of the continuum is a very large gray (or blue) area where a government agency performs some management functions and farmers perform other functions. For purposes of discussion, we can divide the continuum into four types from more to less government involvement:

Type 1: Government does everything. In Malaysia, the Department of Irrigation and Drainage provides for the operation and maintenance of the main and secondary canals, while government sponsored farmers' organizations are responsible for providing water to individual farms. Farmers have no responsibility, and make no management decisions, about the water upstream from their outlets.

Type 2: State dominates; users help. The conventional management division in large irrigation systems is that the state takes responsibility for operation and maintenance of the head works such as a dam or river diversion, and the main, secondary, and larger tertiary canals, while farmers are responsible for managing water distribution and maintenance along the lowest level canals. Typically this entails farmer groups of between 10 and 50 farm families who are expected to work out sharing arrangements on their own.

Type 3: Users dominate; state facilitates. In some countries, associations of water users enter into contractual agreements with state water agencies for the provision of specific water services. In the case of Mexico, the National Water Commission manages the head works and main canals, while legally recognized water user associations employ their own technical staff for the management of the secondary and tertiary levels of the canal networks. Farmers pay their associations for the water, and a small portion of that fee is passed on to the National Water Commission for their services.

Type 4: Farmers do everything: In the Hill regions of Nepal most of the irrigated area is in the hands of local communities who have constructed their own canal systems, generally tapping small stream flows. Similar examples of local, farmer-managed systems can be found in nearly every country where irrigation is important, and the rules and customs of such systems provides a valuable pool of local knowledge that can be tapped in developing new irrigated areas.

Users' Organizations Different than PIM

Management approaches in irrigation generally fall into three categories: (1) public sector management, (2) private sector management, and (3) users' organizations. This last type can be termed "userism," and the process of transferring management to users can be termed "userization". The concept of "userism" is quite different from "privatization" in that we are talking about transferring management not to a third party "owner" who would purchase the irrigation system from the government and then hire out irrigation services to farmers. Rather, the PIM concept is more akin to an employee owned business that gives equal shares. Countries can be ranked according to their level of "userization" in the irrigation sector. While the ranking is rough, the trends are real.

At the upper end of the "userization" graph we find the United States, France, and Japan. Irrigation users have largely replaced the state in managing the irrigation sector although the government retains regulatory functions. At the lower end of the scale, where the state continues to dominate most aspects of irrigation and down to the tertiary or even quaternary levels, we find Morocco, India, Pakistan, etc.

The strategies that countries have taken in implementing PIM policies may be characterized according to three basic approaches: (1) the rapid "big-bang" approach of Mexico where water users are strongly pressured to establish an organization to replace the government, (2) the "bottom-up" slow approach of the Philippines with a strong focus on organizing and consensus building, and (3) a hybrid approach which adopts a moderate pace, such as that adopted by Turkey.

Is participation always necessary? Doesn't participation interfere with efficient management in some circumstances? Do we have to allow farmers to come into our board rooms and advise us on how to do our jobs? Aren't there some natural limits to what irrigation professionals should be responsible for and what farmers should become involved in?

A good rule of thumb is that a participatory dimension is important to all management functions. Perhaps there are exceptions to this general rule, but within the field of irrigation management, it is difficult to imagine any. This does not mean that a farmer's council has to be consulted before any decision is taken. If the water availability is so small that only 40% of the demand can be met along a given canal, do farmers need to be asked if they want the water? However, the farmers who receive only 40% of their demand do need to know about overall water availability so they can plan their response, and perhaps suggest better ways of utilizing their reduced share.

"Userism" as a Management Type

We may broadly classify management relationships into three kinds: the first is public management such as the irrigation department.

The second is private management such as the Continental Corporation which produces Sparkletts mineral water. A third type of management is neither public nor private in the usual sense. We may call this a users' entity, such as a water user association. To describe this type of entity, we may use the term, "user-ism". You will not find this word in any dictionary; it was coined by Mr. Asif Kazi, Special Secretary in Pakistan's Ministry of Water and Power. We have adopted the term because it captures in one word the process of transferring management from the public sector to organizations of users.

Among these three basic types of management, as applied to irrigation systems, the most rare type is private management. This is mainly because irrigation water is a social good involving large numbers of small farmers, and it is very difficult for a commercial company to manage it with profit. This type of management is clearly not a general option for the irrigation sector. What about management by the public sector? While this is the most common type of management that we see today, in most cases public management has low efficiency and requires substantial subsidies. Experience from many sectors, including irrigation, tells us that it is almost impossible to bring public management into high levels of efficiency. Certainly it is possible to improve the management of public irrigation systems, but it is an uphill battle. The interests of the public managers are unlikely to coincide with the interests of the actual users.

The remaining management option is management by users, or participatory irrigation management (PIM). Under this situation, the managers have a direct incentive to manage the irrigation system efficiently because they are themselves users or are directly accountable to the users. This is the logic of userism: we can ensure a coincidence of interests between managers and users because the users are themselves the managers, or the employers of the managers.

Implementation Strategy

The opportunities for participation are different in each phase of the project cycle. Much of the emphasis on PIM has focused on participation in O&M, and particularly in the recovery of O&M service fees on behalf of the irrigation agency. While this aspect of participation is of great practical importance, there are many ways other aspects of irrigation management where participation can be incorporated. These include: (1) participation in irrigation project identification, planning, and design; (2) participation in system layout and construction; and (3) participation in project monitoring and evaluation.

In short, any aspect of irrigation management can have a participatory dimension. We have discussed why participation us important. In this section we will consider how to achieve it: how to implement participatory irrigation management. There is no recipie for this; indeed, the process of formulating a strategy that fits the specific features of any given country is the first — and ongoing — step. There are some common issues to consider, however, of which we will discuss two:

(1) creating an enabling environment, and

(2) start-up, pilots, and expansion phase.

Creating an Enabling Environment:

- Willingness and interest of stakeholders
- Strengthening interest and commitment
- The role of a PIM "promoter"
- Building consensus about the need for change
- Building Consensus among Farmers
- PIM as one part of a broader "package".

Interest and Willingness of Stakeholders: For participation to work, the government, the incumbent power broker and major "stakeholder" in most national irrigation sectors, must be willing and interested. At least three sections of the government must be willing and interested to support PIM: Political leadership, Administrative leadership, Irrigation agency leadership.

A second, and some would argue more powerful stakeholder must also be brought into any discussions of PIM policies at the earliest stage. We refer here to the farmers themselves, who have the most to lose — and also to gain — from changes in the way their irrigation systems are managed. The membership of farmer stakeholders overlaps with governmental stakeholders in the form of political leaders who represent farming constituencies. These political forces who can speak on behalf of both government and farmers, can be particularly important in both designing and promoting PIM reforms.

Strengthening Interest and Commitment: Expressions of interest in PIM on the part of these entities can be strengthened by facilitating participation in workshops where international experiences are shared (e.g., WBI seminars in Mexico, 1994; Turkey, 1996; Tokyo, 1997; IIMI/FAO seminars in China, 1994; Thailand, 1996). Specific study tours to "model" countries and schemes can be arranged. The

World Bank had arranged such study tours to Spain, Mexico, and recently to Argentina and Chile.

Teams of political and administrative leaders have participated in these visits. The first such visit to Mexico by Turkish senior policy-makers was followed by study tours of several contingents of government officials from the Public Works Department. These visits and the shared experiences clearly contributed to the speedy implementation of the PIM program in Turkey. In Egypt, a tour of the USAID-assisted Irrigation Improvement Project was organized for legislators by the Ministry of Water Resources. The tour contributed to a much better understanding of, and subsequent policy support for, shared financing and management of the irrigation system by the government and water users. Methods and tactics for promoting PIM is explored in detail in another section of this Handbook.

The Role of a PIM "Promoter": One clear lesson from participatory experience is the importance of a PIM promoter or "champion" who is effective in mobilizing support within the government/irrigation agency. Typically this role is played by a key official within the agency. In the Philippines, for example, the Assistant Administrator of the NIA is rightfully considered the father of participation in that country's irrigation sector. In Mexico, there were well placed champions within the national water agency (CNA). Perhaps it is possible to rely on a senior consultant (who has the ear of top officials) to mobilize support within the agency. This is the approach being tried in Orissa, India, where a consultant on "farmer organization and turnover" will serve as a guide, but not a direct supervisor, to irrigation department field staff re-trained as organizers. A cadre of specialized social organizers will also be involved as assistants to these engineers. The job profile of the consultant on farmer organization and turnover is given in Annex One.

Building Consensus about the Need for Change: Stimulating a policy dialogue about the need for change and options to be considered can be based both on experience within the country, and by examples from outside the country. Within the country, public discussion of problems faced by farmers whose water services are unreliable or inequitable could be complemented by highlighting other cases where management reforms have resulted in improvements. Universities and centres of higher learning can help organize public discussion events. The media can help in reporting the outcome of these events and highlight key issues in the debate. NGOs can assist in seminars with water users. In Egypt, a professionally-made video film was made available to television stations that attracted large rural audiences in a pilot project area.

Bringing policy makers into contact with PIM cases in other countries is a high priced activity, but can be a powerful ingredient in swaying long-held opinions. Study tours, if carefully arranged and if the right people are involved, can make dramatic differences in the outlooks of individual officials. When Turkey was considering reforms

Building Consensus Among Farmers: A danger in designing and implementing PIM programs (and a cause for their failure) is the lack of attention to farmer interest and support. Since the impetus for PIM oftentimes originates in the perilous state of irrigation agency budgets, the focus for PIM from the agency's perspective could be quite limited, e.g., targeting financial contributions from farmers. The question, of course, is: why should farmers be interested in organizing themselves, if only to pay higher fees? Improvement in services and potential for income enhancement are better motivators for user organization. Farmers' incentives cannot be assumed, but rather must be assessed through field interviews and discussions with a representative sample of the community concerned. Farmers' involvement in designing a management model for PIM which builds-in strong incentives is a critical and often neglected step in the overall PIM implementation process. These issues are explored in another section of this handbook.

PIM as one Part of a Broader: Farmers are interested in much more than just the irrigation system; they want to improve their agricultural production, and more broadly, to improve their livelihoods. In Mexico, when the National Water Commission first met with farmers to discuss their interests, and what their priorities would be for irrigation improvements, the Commission expected that the farmers would ask for more lining of their canals. This is an expensive but important improvement to canal networks in many areas. The Commission was surprised to hear from the farmers that what they were most interested in was full management autonomy. Instead of a physical improvement, they were looking for an institutional improvement. Both types of improvements were within the authority of the government to arrange, but through different means; one required a simple financial outlay; the other required legal changes.

Start-up, Pilots and Expansion Phase

The design of an implementation strategy involves planning for the start-up, piloting and expansion phases of a PIM program:

- *Start-up.* To begin with, management responsibility for PIM must be assigned at the strategic management as well as the operations management levels. At the strategic level, a steering committee could be formed of key officials in the respective ministries, say, public works/irrigation/water resources, agriculture, planning, and finance. This group would approve of the PIM program — its goals, strategies, and a specific work program and budget on an annual basis. At the operations level, responsibility for PIM would be assigned to a specific senior manager. Experiences as to who should be assigned the responsibility vary among countries. In some cases, a senior manager is designated exclusively for Farmers' Organization. In others, a senior manager for Planning or O&M is assigned the additional responsibility. There is no formula for the assignment but whomever is chosen should be interested and committed to PIM ideas. This senior manager would need a group of interested staff from various levels, and representing social science as well as engineering, to help him implement the program.

- The start-up phase would also need resources for preparatory work for the PIM program. In-house expertise may be adequate for this purpose, but often external consultancy inputs may be required. A budget to prepare a pilot three-year program would be a good start and would have to be secured. Sources of financing could be the government's own funds or other external agencies such as bilateral and multilateral donors.

- Piloting. PIM is usually not a new concept to the region or the country. Farmer managed systems exist in most countries. Nevertheless, what is usually new is PIM in all stages of the project cycle, from planning and design to construction and O&M or PIM in the form of irrigation management transfer. In this context, it is useful to talk of "experimenting" with or piloting the PIM idea to test the appropriateness of the various PIM elements to local conditions in the country.

- Yet because PIM implies a new paradigm for the government's relationship with farmers, piloting is not possible in the conventional sense of the term. The irrigation agency must establish a commitment to the basic principles of PIM as a precondition for success in the pilot. The purpose of the pilot should be to help inform the agency how best to operationalize PIM policies, and to demonstrate what kinds of PIM approach would be feasible. It should not be seen as a "wait-and-see" test of whether PIM is a good idea. Inevitable, there will be an internal demonstration function of any PIM pilot, but the management of the agency, at the very least, should support the concepts of PIM at the time the pilot is launched.

- The objectives of piloting are to learn from experiences on a small scale so that a manageable irrigation system or subsystem can be the focus of implementation, monitoring, and learning. The implication is that the pilot projects should cover a range of conditions and be carefully monitored so that changes can be introduced in the original PIM model. For instance, in Egypt, 6 pilots covering about 70000 acres were launched in various parts of the country in the late 1980s. These experiences were evaluated in the early 1990s so that the lessons could be used in another set of irrigation schemes serving a command area of about 250000 acres.

Selection of pilots needs careful consideration. Selection criteria would normally include:

- interested farmers in the area
- relative lack of conflict among farmers or a keen desire to end the conflict
- water delivery system that functions relatively well
- tangible improvements can be demonstrated in a relatively short period of time
- supportive irrigation O&M field staff enjoying good rapport with farmers
- local government officials do not oppose the pilot
- size of the pilot scheme is neither too large (and therefore unmanageable) nor is it too small (results have no visible impact on the rest of the government-managed schemes).

To ensure their relevance, pilots must be part of a phased program that includes start-up, pilots, and expansion. Thus, the Mexican program aimed at the management transfer of all of its large scale irrigation systems (below the secondary canal level) covering nearly 3 million ha in a phased program, learning from experience during implementation. All of the phases should ideally be designed before pilots are implemented, thus placing them in the overall context of a country's institutional reform program. An additional problem with pilots is that they attract a great deal of resources, sometimes disproportionate to the size and scope of the experiment. Replicability becomes a casualty.

Expansion: At the end of the pilot phase, policy makers and managers would have derived lessons regarding the scope for user participation in the management of water, funds, operation and maintenance and capital development of the system. They would also be wiser about current government policies and procedures that help or hinder effective participation. The period before the large-scale expansion phases presents an opportunity to bring about changes in existing laws, policies, financing arrangements and procedures as the need may be. Without this intervention, an expanded program could run into severe problems. What was possible in an informal, small scale setting during the pilot phase may not be easy during the expanded large scale phase.

In addition to required changes in current policies, agency managers have to face the question of institutional capacity — trained staff, budget, managerial resources — to undertake a large scale expanded program that would cover a larger geographical area. Some time may have to be initially spent during the expansion phase in strengthening institutional capacity to manage the effort. These issues will be further explored in the section on Training Strategies.

Training Strategy

What Training do the Organizers Need?

Assuming that agency staff will be used for the organizing, and assuming that their professional training has been in irrigation engineering, they would need to be re-tooled as organizers. First of all they would need a thorough understanding of the rationale for PIM, so that they can present a clear message both to the farmers and to their own colleagues within the irrigation agency.

Secondly, they will need training in communication skills (including listening skills) for effective interaction with the users. Thirdly, they will need training in social analysis, including an understanding of social stratification (by caste, ethnicity, or class), kinship, patron-client relations, labour relations, religious factors, political affiliation, land tenure (tenants, share-croppers, owners), etc. And fourthly, they would need training in methods for gathering information from farmers (e.g., participatory rural appraisal) and in methods for organizing farmers.

Legal Framework

The legal framework for the establishment of WUAs, and for enabling them to operate and maintain such parts of the irrigation system, consists basically of three sets of legal instruments, namely:

- The enabling law,
- The bylaws of the WUA, and
- The transfer agreement between the irrigation agency and the WUA.

The existence of a basic law on water, or on WUAs, is certainly an important parameter for the other parts of the legal framework because the basic law would usually specify the main issues that need to be included in the bylaws and transfer agreements, and would also determine the manner in which those issues are addressed. Those issues would usually include the procedure for establishing WUAs, the rights and duties of the WUA and the irrigation agency and the relationship between them, and the structure of the water rates and other fees.

The Enabling Law

For a WUA to be established as a legal entity, there has to be a law authorizing its establishment. This law could be a general comprehensive "Water Law" that deals with all aspects related to water, including establishment of WUAs. The National Water Act in Mexico, and the Water Resources Act in Nepal, are examples of such a comprehensive law. The enabling law could also be special rules and regulations dealing specifically with WUAs, and deriving their authority from a basic law, such as the "Implementing Rules and Regulations on the Provisions of Republic Act No. 7607" on small farmers in the Philippines.

Because of the absence of a basic law specifically on water or on WUAs, the states of India have relied on different laws to establish WUAs. In the Indian State of Maharashtra, WUAs have been established and registered as co-operative societies under the "Co-operative Societies Act." On the other hand, in the State of Tamil Nadu and the State of Orissa in India, WUAs are established and registered as societies, under the "Societies Registration Act."

The law establishing WUAs would usually include provisions indicating that the WUA to be established is a legal entity. Such enabling law would also address the relationship between the WUAs and the irrigation agency, the duties and obligations of the irrigation agency, and those of the WUAs, and the structure of water rates and the operation and maintenance and other fees. The enabling law may also lay down some of the main issues to be addressed in the bylaws of the WUA, and in the transfer agreement.

All societies, including WUAs, registered in any state in India under either the Co-operative Societies Act, or the Societies Registration Act, are legal entities capable of contracting, opening and operating bank accounts, and instituting and answering suits. However, because of their general nature, those two Acts do not address certain WUA related issues that are usually addressed in special water law or through a law establishing WUAs. It is for this reason that specific legislation dealing specifically with WUAs is often desirable.

Bylaws of the Water User Association

Whether established under a separate law or under an umbrella enabling law, the WUA would normally be required to prepare and agree on its bylaws before it can be registered as a legal entity. Those bylaws may be called "Regulations," "Constitution," "Charter" or "Articles of Associations." The issues that such bylaws need to address include:

Basic Facts About, and Objectives of, the WUA: The basic facts would include the name of the WUA, the law under which it is registered and its registration number, its address, and a clear definition of the area that the WUA is serving or its area of operation. This area of operation could be an entire irrigation district, or an entire command of a distributary, minor, sub-minor or a water course. It could also be defined by its size in acres or hectares. A broad statement on the objectives of the WUA is usually included in the bylaws. Such objectives would include: participation in the management, operation, maintenance and upgrading of the irrigation infrastructure works that the WUA has taken responsibility for, collection of water charges, and provision of irrigation and drainage services to the members of the WUA.

Criteria for Becoming a Member of the WUA: Most bylaws restrict membership of the WUA to the registered land owners in the hydraulic unit, who are engaged on a full-time basis in farming. If any member of the WUA sells his land, his membership will be automatically cancelled, and the new owner will be eligible for the membership of the WUA. However, the bylaws in some countries extend the right to become a member to both owners and tenants. In the Indian State of Maharashtra membership of the WUA is extended to "Any owner/cultivator/permanent tenant/protected tenant in the area of operation of the society...." These bylaws extend membership of the WUA beyond owners and tenants to include other categories of users such as sharecroppers and encroachers who are prevalent in some parts of India.

Number of Farmers Required for the Establishment of a WUA: Most of the bylaws state that at least 51% of the registered land owners in the command area where the WUA would be established should be enrolled as members before the WUA can seek registration, and before it can be allowed to operate. However, some bylaws allow the WUA to seek registration based either on the number of farmers enrolled, or the size of the land holdings coming under its operation regardless of the number of farmers enrolled.

In Orissa "At least 51% of the registered land owners in the command area covered by the Association should be enrolled as members, or the land holdings of the members should cover at least 51% of the total area under the proposed Association." Moreover, in some cases, once 51% of the farmers are enrolled and registered in the WUA, all other farmers within the command area will be deemed to have become members in that WUA. In Mexico "all users listed in the register who, while not founder members of the Association, apply and pay for irrigation services, thereby tacitly agreeing to belong to the Association, shall also be deemed to be members with the same rights and obligations."

The WUA as a Legal Entity: Although the enabling law would usually specify that the WUA is a legal entity, further details regarding what this entails are usually included in the bylaws. It is usually stated that the WUA is authorized to enter into contracts in its name, and that the WUA can sue in its own name, and answer suits instituted against it. The WUA can also be authorized to borrow funds from private sources, using, if necessary, its assets as a collateral.

Structural Organization and Internal Management: Although WUAs may be organized differently, the two most common ways of organizing a WUA at the membership level are:

(i) a general body of the WUA which would consist of all registered members who are current in the payment of their dues, as in Mexico, Nepal, and the Indian states of Maharashtra, Orissa and Tamil Nadu; or

(ii) a general body, which could be called the general assembly, which would consist of delegates directly elected to represent the different irrigation districts or sub-units within the hydraulic unit, as in the case of some WUAs in Turkey. In the latter case, there is an absence of one single forum encompassing all members.

The executive body would usually consist of a president (or chairman), a vice president (or vice chairman), a secretary, a treasurer, and other specified number of members, and those posts do not, in many countries, carry remuneration. One exception is Turkey where the president of the WUA is paid a salary, and the other members of the executive body are paid honoraria.

Membership of the executive body may also be extended to nongovernmental organizations (NGOs) or any other institutions interested in irrigated agriculture, including a representative of the irrigation agency, but such members usually do not have the right to vote.

A bank account in the name of the WUA is usually opened, with separate sub-accounts for water charges, operation and maintenance fund and membership fees. The executive body would also approve work expenses, engage labour, organize labour contributions from members, keep systematic accounts and records of amounts collected and those spent on the water charges, operations and maintenance fund, and membership fees. The executive body is usually required to have such accounts and records audited annually, and submit such accounts and audit reports to the general body during the annual meeting for approval. The executive body could include members who are specifically designated as auditors.

Operation and Maintenance: One of the primary objectives of the WUA is to operate and maintain the transferred irrigation and drainage system efficiently and economically, and with the full and active participation of all the members. Operation would include receiving water in bulk from the irrigation agency at a prescribed rate at the head of the minor/distributary and distributing such water equitably and in a timely manner, as per procedures and criteria agreed with the irrigation agency, to all farmers in the hydraulic unit, whether members or non-members. The bylaws would lay down, in agreement with the irrigation agency, the criteria for allocation of water to both members and non-members, which could be based on the type of crop grown or the size of the area to be irrigated, or both. The bylaws would also include the criteria for assessing water charges and operation and maintenance charges from both members and non-members. The operation and maintenance fund could include sources other than charges from farmers.

Water Charges: Reference to water charges is usually included in the enabling law. Other details on water charges are also included in both the bylaws and the transfer agreements. Inclusion of provisions on water charges in the bylaws would serve the purpose of establishing the payment obligations of each member of the WUA, whereas the provisions in the transfer agreement would establish the payment obligations of the WUA, as a legal entity, vis-a-vis the irrigation agency.

The transfer agreement would usually include provisions on the manner in which water charges are calculated, and the due date or dates for payment of the water charges by the WUA to the irrigation agency. It may include provisions for the payment of commission or discount by the irrigation agency to the WUA on water charges collected by the WUA. The transfer agreement would also include provisions giving the irrigation agency the right to suspend delivery of water if the WUA fails to make the payments of the water charges within the prescribed or extended time limit. Moreover, non-members of the WUA may be required to pay higher water rates than those paid by members.

Rights and Obligations of Members: Under most WUAs, each member of the WUA has one vote regardless of the size of his land holding. The bylaws would also need to address the issue of proxy voting-whether it is allowed, and if so, the maximum number of proxy votes one member may cast on behalf of other members. Failure of a member of the WUA to meet his membership obligations as described in the bylaws, such as failure to make payments, permit inspection of the irrigation system in his land, comply with the terms of the transfer agreement, carry out proper maintenance, or allow delivery of water to other users may subject such a member to sanctions. Such sanctions may include suspension of such a member until all outstanding obligations are met.

Interpreting and Amending the Bylaws: Provisions would usually be included in the bylaws themselves describing the procedures for interpreting provisions of the bylaws in case there are different views as to what a certain provision may mean. Procedures and quorum required for amending the bylaws would also be included in the bylaws. Usually interpretation of provisions of the bylaws would be referred to a central body such as the Registrar of Societies, or the irrigation agency, and amendments would be effective after approval of such body.

Establishment of a Federation of WUAs: The bylaws of some of the WUAs, as in Mexico, Colombia, Tamil Nadu and Orissa, include provisions enabling the establishment of a federation encompassing registered WUAs in one command area. Those bylaws specify the responsibilities of the federation, and the relationship between such a federation and each of the affiliated or member WUAs. Usually, the presidents of each of the WUAs that decided to establish or join such a federation would represent their WUA in that federation. The federation would have an advisory non-binding role over the member WUAs, and may be used to resolve any disputes among such member WUAs, or between WUAs and the irrigation agency.

Bylaws of the Water User Association

The transfer agreement is the agreement between the WUA and the irrigation agency in which the irrigation agency agrees to transfer to the WUA responsibilities for operation and maintenance of certain parts of the irrigation system, including the drainage system, and the collection and remitting of water charges; and the WUA agrees to carry out such responsibilities. This agreement may also be called "Memorandum of Understanding (MOU)," "Transfer Protocol," "Concession Agreement" or just "Concession." The issues that such transfer agreement would need to address include:

Area and Irrigation System to be Transferred: The agreement would need to define clearly the irrigated area to be transferred, specifying the size of the area, and the command under which it falls, and including the irrigation system existing there that is being transferred. The system to be transferred is usually the irrigation system at the level of primary, secondary and tertiary, including the drainage of such areas too. A copy of a map showing such area may be attached to the agreement. The agreement would specify whether the ownership of the irrigation system, including the land and structures and works there on, remains with the irrigation agency or is being transferred to the WUA, together with the operation and maintenance of such irrigation system. Provisions should also be included clarifying whether the ownership of any ancillary equipment is being transferred.

Interim Joint Management: Some agreements may provide for a joint management of the irrigation system for a short period of time by both the irrigation agency and the WUA. The rationale for such joint management is to prepare the WUA, during this interim period, for taking over full responsibility for operation and maintenance of such irrigation system. During this interim period which may run for up to one year, officers of the irrigation agency would train WUA representatives in compiling necessary data, and preparing and testing operation and maintenance plans for the distributaries, minors and subminors to be transferred to them. They would hold joint inspections to identify any problems in the irrigation and drainage system, and to agree on how to deal with them.

Transfer of the Irrigation System: Transfer of the irrigation system to the WUA will normally be preceded by a number of actions, including the preparation of an inventory of the works, structures and equipment to be transferred, joint inspections of those works, structures and equipment by both the irrigation agency and the WUA representatives, carrying out of necessary testing, and repairs, if any, at the irrigation agency cost, and handing over management of the system, along with all necessary documents and instructions, to the WUA. The WUA would need to satisfy itself that the system is, indeed, in a good working condition, as the transfer agreement would include provisions that the irrigation system, at the time it is transferred, was in good working condition.

Responsibilities of the Irrigation Agency: The agreement would spell out clearly the responsibilities of both the irrigation agency and the WUA. Responsibilities of the irrigation agency would include handing over the system in a reasonably operating manner, delivery of water to the WUA in bulk at the agreed time and providing the WUA with any agreed upon financial assistance and other benefits. The agreement could include provisions absolving the irrigation agency from liability should it be unable to deliver the agreed upon amount of water, or unable to deliver it at the agreed time, for reasons of force major, or act of God. In periods of water scarcity or emergency, after the demand for domestic and other priority uses is satisfied, the irrigation agency would usually have the authority to decide that the remaining water for irrigation shall be allocated to crops of utmost importance to the community there.

Responsibilities of the WUA: A number of the responsibilities of the WUA detailed in this section of the agreement are usually spelled out in the bylaws of the WUA, but may still be included in the transfer agreement to clarify the obligations of the WUA towards the irrigation agency. Such responsibilities would include: operating and maintaining the irrigation system transferred to it, including the drainage system, in a proper and satisfactory manner; receiving water in volumetric basis, and distributing such water equitably and in a timely manner, based on clearly defined criteria, to both members and non-member farmers in the operation area, and collecting the water charges agreed with the irrigation agency.

The responsibilities also include establishing the operation and maintenance fund, and maintaining and repairing, in a satisfactory manner, all the field channels, field drains, minors, subminors and distributaries, together with the structures there on in the operation area of the WUA. In addition, the WUA may be responsible for the maintenance and repairs of any equipment and machinery transferred to it. Such equipment and machinery may be transferred to the WUA as part of the irrigation system for which it is now responsible, or may be separately leased by the irrigation agency to the WUA at an extra cost.

The WUA would also be responsible for the security of the infrastructure transferred to it, and such responsibility could either be carried out by the members of the WUA themselves, or through hired labour. Maintenance would usually include silt clearance and removal of weeds from all water courses under the WUA. It would also include earthwork to restore banks and repairs to other structures, in addition to maintenance of service roads.

Usually minor repairs are carried out by the WUA and major repairs by the irrigation agency. Definitions of what is "minor" and what is "major" should be included in the transfer agreement (repairs to damage caused by natural disasters such as heavy rains, floods or earthquake are usually considered major repairs). The WUA may be required to prepare an annual maintenance program for the irrigation system under its responsibility, including any machinery and equipment, and to submit such program to the irrigation agency for approval prior to implementation. Operation and maintenance of the irrigation and drainage system, other than the one transferred to the WUA, would continue to be the responsibility of the irrigation agency.

The agreement would authorize the irrigation agency to suspend supply of water to the WUA if maintenance and repairs were not being carried out properly, or to carry out the repairs itself and recover the cost from the WUA. The agreement would also include provisions on how disputes between the WUA and the irrigation agency, arising in the course of the operation and maintenance of the transferred irrigation system, would be settled. Such disputes could be referred to a committee comprising one representative from the irrigation agency and the water users' federation.

Termination of the Transfer Agreement: Although the enabling law may include provisions on the termination of the transfer agreement, usually more detailed provisions are included in the transfer agreement itself. The agreement terminates after expiry of the number of years specified in the transfer agreement, which may be as high as twenty years as in Mexico. However, the agreement would usually be subject to renewal for another similar period. Moreover, failure by the WUA to comply with the provisions of the agreement, including the failure to properly operate and maintain the irrigation system transferred to it, or to make timely payment of water charges, or to take corrective measures within a specified period of time, as agreed with the irrigation agency, would give the irrigation agency the right to terminate the transfer agreement.

Organizing Processes

PIM implies the establishment of an organization of water users. In cases of management transfer, the establishment process focuses on a level of user organization where no organization presently exists, that is, at a level of the irrigation network previously managed by the government irrigation agency. Typically this entails organizing federations of user groups at the tertiary level under a secondary-canal association. The challenge of creating a new organization of users is perhaps the most central feature of the management transfer process. The act of management transfer from the agency to the users depends upon a user organization that is capable of assuming those management responsibilities.

Before any organizing of the users is carried out, there needs to be a package of incentives in place for both the users and the agency staff whose jobs would be affected by the transfer program. Such incentives are needed both to make the program work, and to maintain credibility with farmers whose long-term support will be required. If organizing is attempted before an adequate incentive structure is established, the transfer program could well collapse, thus setting the entire program back by several years, as well as causing short-term hardships to those concerned. Thus, if the incentives are not clear and attractive to farmers, the organizing process should be delayed until the incentives are clarified.

First determine:

- Who will do the organizing?
- What training do the organizers need?
- Who will provide the training?

Then, the organizing steps for transfer of management to users.

Who will do the Organizing?

The first step is to decide on the type of organizers who will work directly with farmers in helping establish the organization. [We use the term "water user association" (WUA) to refer to this new organization, although the term, "WUA" can also refer to organizations at the tertiary level of the system which have always been outside the management control of the government irrigation agency.] In the Philippines, a special cadre of social organizers was recruited and trained by the National Irrigation Administration in the late 1970s and early 80s. These organizers were mostly social workers or social scientists selected for their ability to work easily with farmers in village conditions. They were trained in irrigation management so they could better understand the technical problems of the farmers they were trying to organize.

The work of the Philippines social organizers was effective, yet this approach has not been widely replicated by irrigation agencies in other countries. The investment and bureaucratic difficulties involved in recruiting new temporary staff from a different discipline has led many irrigation agencies to try other approaches. In Mexico's transfer program, the National Water Commission's own staff were used along with staff from a sister agency, the Institute for Water Technology (IMTA). In addition, a few consultants were brought in on a case by case basis. In India, some state irrigation departments have relied on extensionists from the Agriculture Department under the Command Area Development Program. Also in India, several NGOs have been involved in organizing, at the request of the irrigation departments. But the organizers of choice, for most irrigation agencies, will be their own field staff. These staff are already within the bureaucratic structure of the agency, so the lines of authority are clear, there is little additional expense involved, and these staff are already familiar with the physical systems and with the local farmers.

The problems with using irrigation agency field staff for organizing work, however, are considerable. These staff are not necessarily interested in organizing farmers, They have not received any prior training in social work (in most cases), and their superiors have also no training (nor interest) in these tasks. These would-be organizers must be re-trained for their organizing tasks, and just as importantly, their job assignments need to be redefined to reflect their new role. In addition, their superiors need to be trained and re-oriented so they understand and appreciate the new role to be played by their field staff.

What Training do the Organizers Need?

Assuming that agency staff will be used for the organizing, and assuming that their professional training has been in irrigation engineering, they would need to be re-tooled as organizers. First of all they would need a thorough understanding of the rationale for PIM, so that they can present a clear message both to the farmers and to their own colleagues within the irrigation agency.

Secondly, they will need training in communication skills (including listening skills) for effective interaction with the users. Thirdly, they will need training in social analysis, including an understanding of social stratification (by caste, ethnicity, or class), kinship, patron-client relations, labour relations, religious factors, political affiliation, land tenure (tenants, share-croppers, owners), etc., and fourthly, they would need training in methods for gathering information from farmers (e.g., participatory rural appraisal) and in methods for organizing farmers.

Organizing Steps

Note: This assumes there are reasonable incentives for farmers to take over system, and for the government irrigation agency to hand over the system; if not, go back and work on the incentives!

- Organize the organizers: Arrange for supervision/support for the organizers, and clear lines of communication with Departmental staff responsible for the overall project (Ensure that PIM component of program is well integrated with rest of project).
- Meet the farmers and other irrigation stakeholders/Discuss plans formulated during participatory design phase:
  * village head
  * local administrative officials
  * local political leaders (MLAs/MPs)
  * leaders of other farmer organizations (producer cooperatives).
- Identify key power relations among farmers; develop strategy for organizing
- Establish provisional boundaries of the system (through consultation with key power brokers among the farmers); conduct inventory of potential members; draw map showing command area and irrigation system.
- Arrange series of meetings between farmers and Departmental field staff to discuss improvements that need to be made prior to handover.

- — arrange canal walk-through to discuss specifics of design/ infrastructure improvements
- — discuss general terms of WUA contracts.
- Arrange farmer visits to other associations to discuss with those farmers (and invite those farmers to visit new association).
- Organizational Assistance:
  - — Help prospective WUA leaders arrange farmer meetings to discuss plans
  - — Help formulate/revise bylaws
  - — Advise on elections/selections
  - — Assist with legal registration
  - — Arrange management training for WUA leaders.
- Arrange meetings between WUA leaders and Department staff to discuss details of WUA contract and terms of transition phase leading to hand-over;
- Advise on staff recruitment(?)
- Assist with formal hand-over
- Visit periodically to monitor WUA's performance.

Reorienting Agencies

Agencies implementing PIM need to reorient themselves to a new style of water and infrastructure management. No longer is the agency the sole manager of the water system; rather, the agency staff become management partners with farmers. The agency retains control over the water supply and head works (in most cases) and perhaps the main canal network. Their management must be closely coordinated with the management of the lower levels of the system, now under the control of user associations. Under this new arrangement, the agency has both a direct management role (for the highest levels of the system) and an indirect management role — facilitating and supporting the work of the user associations.

- Restructuring the Irrigation Agency
- How agencies do NOT Change under PIM
- Facilitating and Hindering Factors.

Restructuring the Irrigation Agency

Even without an explicit PIM program, changes over the past two decades in the irrigation and water policy context clearly point to needed changes in agency roles and functions. Irrigation agencies are being forced — by citizen concern as well as by the conditions imposed by international financial organizations — to pay a great deal of attention to social and environmental impacts of irrigation activities. These have become important aspects of project assessments in the 1980s. Similarly, as user participation has gained importance, irrigation agencies have had to re-evaluate their roles and functions with respect to irrigation development and management.

Incorporating PIM requires a range of new functions and new organizational structures for the irrigation agency; it must shrink and shift to make room for user management. It must shrink in staff and functions, and it must shift from directly doing O&M to indirectly supporting O&M within the user managed level of the system. Specifically, changes in agency roles require the agency to perform new tasks as explained below.

- *Revised Structure:* At least some field-level staff of the agency, and perhaps most field staff, would be replaced by the new staff of the WUAs. As a general rule, higher level staff of the agency would be less affected by the transfer of O&M responsibilities.

- *Consultation:* Staff in the various functional departments of the agency need to commit to and gain expertise in farmer consultation.

- For those staff in planning and design, PIM implies a consultative step that would provide farmer inputs into prioritising system improvements, selecting designs and layouts, and determining outlet locations. On the part of agency staff and their consultants, consultation with farmers for planning and design requires listening skills and an ability to present technical details in simple language. It also implies the flexibility to change plans and designs based on farmers' suggestions.

- For O&M staff, PIM implies the ability to respond to feedback on system performance from water users. Field staff would need to reorient themselves from carrying on as operating engineers to performing advisory functions — a role in which they may have little experience or expertise.

- *Accountability:* The accountability of irrigation staff increases considerably with PIM and the formation of WUAs. Both the agency and the WUA need to maintain accurate records of water delivery (planned and actual) to ensure that the agreed amount of water is delivered/received. Where PIM is introduced at the survey and design stage itself, information on initial designs and comparative costs would have to be shared with water users. The importance of shared information increases significantly with PIM, since now agency staff are not only providing information "upwards" to their bosses but also "downward' to their clients, the managers of the WUAs.

- *Advisory Services:* Agency staff need to reorient themselves to extend technical and managerial advice to water users. PIM is effective only when informed water users participate knowledgeably in the planning, design, construction and management of irrigation systems. In Egypt, where water users are organizing themselves around a common pump at the head of the tertiary canal, the Ministry of Public Works and Water Resources has established an Irrigation Advisory Service to extend assistance to WUAs.

How Agencies do not Change Under PIM

Many of the irrigation agency's core functions continue after PIM is introduced. In most cases of PIM, the public irrigation agency maintains the legal responsibility for overseeing the wise use of the irrigation facilities. Under most contracts, the WUA is managing these facilities for a set period of time (25 year concessions in Mexico). Cases of outright privatisation — where the government agency permanently sells or cedes the ownership of the infrastructure to a WUA — are very much the exception. Thus, the public irrigation agency has not only a right but an active responsibility to regulate, for example, groundwater levels and soil and water quality.

Management functions also continue, with the agency maintaining responsibility for the head works and (usually) main system. The agency also has a monitoring role in ensuring that the WUA is adhering to the contractual agreement of management transfer, and is properly maintaining the infrastructure and adequately delivering water to the WUA members.

Facilitating and Hindering Factors

Incentives within the agency are often opposed to the establishment of strong WUAs. Organized users pose a threat to the agency's control and to the endemic problem (in many countries) of bribes and kick-backs on contracts. Management transfer programs which aim to expand the role of WUAs threaten the jobs of agency staff. And the task of mobilizing the formation of WUAs becomes an added responsibility for these same staff, and not a task in which they have either professional training or interest. Finally, because irrigation agencies are engineering organizations, the professional rewards and recognition are in design or construction of physical schemes, not in dealing with farmers' demands.

Given this situation, which naturally varies considerably from country to country, how could PIM possibly gain a foothold? How can an agency transform its own management culture? There are a number of tools for doing this:

- *Performance Evaluation Criteria:* Including work with WUAs in the job descriptions and performance evaluations of agency staff is one way to create incentives for PIM. Supervisors (whose own job descriptions also need to incorporate PIM criteria) need to monitor their subordinates' performance in working with WUAs, and request feedback from farmers on how helpful staff members are. Publicizing the performance plans of agencies among users may strengthen agency and staff accountability. While the number of associations registered is relatively easy to measure, it is not an adequate indicator of staff performance with WUAs because it does not take into consideration how well the WUAs and joint management work. Fee collection rates from farmers or from WUAs may provide a better indicator of farmer satisfaction.

- *Monetary and Non-monetary Rewards:* Monetary rewards are often difficult in the context of civil service regulations. In some cases, where the agency is an autonomous corporation, a reward system may be possible. The National Irrigation Administration, a semi-autonomous agency in the Philippines, was able to offer higher than regular civil service salaries to its staff. However, this was contingent upon maintaining a balance between expenses and cost recovery, which gave staff an incentive to work with WUAs to reduce agency costs and raise irrigation service fees. Non-monetary rewards can also provide a powerful motivator for agency staff. In Nepal, a special medal was awarded to an Irrigation Department engineer who had made extraordinary efforts to establish a WUA. Some already motivated staff will derive job satisfaction from challenging new tasks such as WUA organization, provision of technical assistance, monitoring and regulation, rather than routine O&M. But the agency's management must provide clear and consistent signals that such new roles are valued.

- *Reduced Transaction Costs:* Reduction in the number of conflict cases (or "hassle factors") to be resolved by the agency staff has also proved to be a valuable incentive for working with WUAs. The establishment of WUAs as an organized forum for communication can reduce the transactions costs for agencies and for farmers. Reduction in damage to structures, as farmers focus on protection of system facilities eases the burden placed on agency field staff. Thus staff may have to devote less time to field inspections and user management often results in greater vigilance over infrastructure and fewer complaints. Thus, there may be less need to deal with individual farmer demands as WUAs take on additional roles. In Chile, problems with the state management of irrigation are no longer "politicized" by farmers since users' associations and organized forums for articulation of problems have been established.

- *Skills Training:* Preparation is needed for two types of activities that could result from the introduction of PIM:

- Activities dealing with user participation, including farmer organization, WUA establishment, technical assistance for WUAs, fostering agreements with informal and formal WUAs, public communication and information etc.

- Activities that can be taken up by agency staff as a result of water users assuming a larger role in irrigation O&M thus freeing up time and resources for the agency.

Agency management will have to formulate a program of short and long term training to prepare the staff for changes in roles and functions and upgradation of expertise. It may not be possible to meet all needs through training of available staff. New staff may have to be inducted into the agency. Secondment from other departments of government, from universities or NGOs may also be considered. For instance, in Egypt, the Ministry of Public Works obtains the services of agricultural extension staff through secondment from the Ministry of Agriculture. In Philippines and in India, community organizers are hired as consultants and contract staff.

Financial Aspects

O&M can be financed directly (through a direct tax based on O&M costs) or indirectly through land taxes or other kinds of agricultural taxes. The relative share of government vs. farmer contribution varies dramatically across countries, with some countries using O&M as a vehicle for delivering subsidies to irrigated farmers.

- How to set O&M cost
- User participation in defining service levels
- Assessing users' willingness and capacity to pay
- Determine appropriate charging mechanisms
- Linking fees and services.

How to Set O&M Cost

The managing agency needs to identify the various services it provides through the water delivery and drainage systems under its management control and then allocate costs to each service category. This information should be made freely available to farmers, and to their WUA leaders. Such information helps WUAs understand the respective contribution of government and farmers and helps the WUA determine which functions it might take over from government, and which functions it might prefer to pay government to provide. An aspect of the cost determination activity is the separation of "staff or administrative" costs and "works or program" costs. A clear rationale for the two types of costs is important in justifying the irrigation service fees which farmers are expected to pay. This is equally true for the water fee payable to the government agency, and to the fee payable to the WUA.

User Participation in Defining Service Levels

Involvement of farmers/WUAs early in project design to determine the level of service required by farmers is crucial. Users' involvement in setting service levels will ensure the genuine demand for those services, which will be expressed by a willingness to pay the fee required to support that level of service. Some examples of users working with the agency in determining what services the government should provide are given in the following:

- In Mexico, following the transfer of management of irrigation districts to WUAs, irrigation plans are prepared by WUA-hired managers at the level of the module, each covering about 5-8000 ha, taking into account cropping plans, conveyance losses, and equitable distribution. These are then negotiated with the national irrigation agency to determine the final allocation, generally based on an arranged demand pattern. O&M costs at the level of the secondary and below are met by user fees managed by the WUOs. In addition, WUOs contribute to a part of O&M costs of higher levels of the system.

- In the Office du Niger, Mali, joint committees have been established for O&M in every region. The committees have 5 to 10 representatives of producers and 5 to 10 representatives of the Office. These committees decide on types of services, costs including procurement matters, and water service fees. They also make decisions on the use of 50 per cent of the user fees collected for O&M.

- In Chile, the national federation of WUOs was consulted in the design of a Bank supported irrigation project. The federation and local WUOs played an active role in project preparation especially in discussions of service options and costs. Subsequently, the project incorporated the condition of WUO approval for investment proposals and other project components.

Assessing Users' Willingness and Capacity to Pay

The readiness of the user to pay will depend not only on the benefits to be derived, but on the level of payment to which the users have become accustom. It is in the farmers' interest to pay as little as possible. But it is also in the farmers interest to receive reliable supplies of water, and for this he will certainly be willing to pay something. In a rehabilitation project in Egypt which is introducing PIM, it has been estimated that users will derive incremental income from the project due to expansion of irrigated area and yield increases and that incremental costs of their participation in irrigation management would amount to about 20 per cent of this new income. Increases in income would also depend on agricultural extension services and these would have to be strengthened to allow farmers derive full benefits from the improved water services delivered by the project.

Financial viability of the WUA enterprise is critical to the sustainability of the organizations and the irrigation infrastructure. It is important to examine the total costs the WUAs have to bear, including staff, materials, travel to meet with government officials, and formal and informal payments that must be paid to government agencies. If the level of fees members must pay to meet these costs (which are often in addition to continued payments to the government) is too high a portion of gross or net income from irrigation, the WUAs are not likely to succeed. This is particularly problematic for pump irrigation systems, in which high energy and maintenance costs exceed what the organizations are able to collect from farmers.

In some cases, WUAs must depend on income from other sources to subsidize their irrigation activities. For instance, some irrigation districts in the Western United States rely on revenue from power sales to balance their irrigation budgets.

Determine Appropriate Charging Mechanisms

There are a number of alternative charging mechanisms available to agencies and WUOs in structuring O&M fees. As a principle, the fee structure should be equitable, administratively simple, and easily understood by both users of a particular service and the staff that will administer and collect the fees.

User Fees: Generally, user fees are charged by area irrigated. The advantage of this method is that it is relatively simple to understand. However, it does not reflect the quantity of water used nor does it provide an incentive to conserve water. Volumetric charges address those concerns. However, there are many measurement difficulties in operationalizing this method in the field. Agencies often favour a combination of the two methods. In Mexico, for instance, the agency charges the WUAs volumetrically at the turnout of the secondary canal. In turn, the WUAs base their water charges to individual members on area irrigated and type of crop. In addition to unit area and volume, WUAs have also resorted to other bases such as charges for the entire season which clearly favour the high volume user. The design of irrigation fees should be a topic in WUA training programs so farmers understand the implications of various modes of charging user fees for efficiency and equity.

Property Tax: This is a charge based on the value of the land receiving the service. Irrigation and drainage enhance land value. The higher the land valuation, greater the service payments. The problem with this method is that generally land taxes are collected by the Treasury and are not transparently linked with irrigation services. Also, the valuation and revaluation process demands a great deal of administrative resources.

In-kind Contributions: Contributions of labour, materials or both by the benefiting farmers towards O&M reduce the cost of services. In Viet Nam, the provincial and national governments finance schemes down to 150 ha for new irrigation development. Below this level, farmers must build the channels, with the government providing support for survey and design, as well as materials in some instances. After schemes are completed and taken over by farmers, each member must provide up to 20 person days per year towards maintenance of the tertiary as well as secondary systems.

Replacement of Assets: A significant cost in irrigation services is depreciation of the asset base used to store, transfer and deliver water services to users. Facilities employed range from minor tools and equipment, buildings and housing, motor vehicles and heavy equipment, canals and drains, pumping stations, and where applicable, major structures including dams. Unfortunately, depreciation costs are usually neglected in estimates of cost of water services whether provided by irrigation agencies or by WUAs. Assets are consumed at different rates and each year, a significant non-cash value is "lost" as assets depreciate, even with adequate maintenance. Without proper maintenance, service life will be shortened and depreciation will accelerate. The agreement between the irrigation agency and the WUA includes O&M standards to be maintained. Provision is also made for collecting and retaining disaster funds by WUAs. In Mexico, the Hydraulics Committee at the district level approves the annual O&M program proposed by the WUA.

Care of assets is vital to the sustainability of water system performance. Maintenance and the ultimate replacement programs for all assets need to be planned and implemented in an effective and economical manner. In parts of Vietnam, the estimate of costs of services (and hence of user fees) include provision for depreciation of assets. Adequate data about the location, age, condition and serviceability of all assets is therefore essential to this process. WUAs need to be trained and assisted in preparation of an asset inventory and in the formulation of a maintenance program.

Linking Fees and Services

An obstacle to user contributions is the perception that their payments are not linked to the irrigation agency or its services. Often, money for services is collected by the central revenue department, and from there funds are allocated in the various government agencies, including irrigation, with little link between revenue earned and costs of service provision. To reverse this process, the money cycle must be transparent.

Ideally, WUAs would collect O&M fees and manage the system under their control and pay a service fee to the agency for that portion of the system under the agency's control. Under this arrangement, the agency and the WUA agree on mutual rights and responsibilities and the WUA is completely responsible for management of water, infrastructure, and finances for a part of the irrigation infrastructure.

If users are to pay service fees with the government managing the system, fee collection is preferably done through the WUAs. The Philippines experience has shown that fee collection is far better in systems where WUAs are organized and where they have a role and incentive for collection.

For instance, in the pilot projects in Maharashtra, India, WUOs are allowed to retain a proportion of collections as a bonus. The cost of collection by WUAs is lower relative to government agents. Sanctions for non-payment by individuals are easier when enforced by WUAs. The danger is that considerable time and energy are devoted to fee collection on the part of the agency staff as well of the WUAs and fees rather than reliable water services becomes the focus of management attention.

Where users pay service fees to the managing agency, fees are a good means of signalling satisfaction with services, as in the case of Indonesia. Such information is publicized and used for evaluation of agency services and management. In any case, transparent accounts and audits are fundamental in showing WUA members that the financial management of the organization is sound. General body meetings of the WUAs must discuss financial performance including the accounts and audits.

WBI Training Programme

In most developing countries, irrigation development projects and their operation and management are heavily dominated by the public sector. Conventional wisdom has assumed that only the state was capable of handling large modern projects requiring heavy capital investment, complicated technical inputs, the legal mandate to distribute water, and collect fees.

Recent experience challenges these assumptions. Government-operated irrigation systems are often poorly maintained with steadily deteriorating infrastructure. Yet some of these same systems show dramatic improvement when their management is transferred to water user associations (WUAs), which enter into contracts with the government for operating and maintaining portions of the system or in some cases entire systems.

Since the mid-1980s, countries including Mexico, Turkey, Indonesia, Philippines, Colombia, India, Srilanka, and Nepal have adopted policies to encourage greater management participation by water users. One of the most dramatic management transfer programs has been in Mexico, where the government adopted a policy to gradually transfer all its large scale irrigation districts to 78 WUAs. As of mid-1993, the management of more than 1.2 million hectares (ha) of irrigated land has been transferred to WUAs.

In response to interest in participatory irrigation management expressed both within the World Bank and in a number of borrowing countries, the World Bank Institute of the World Bank has initiated a five-phase program on participatory irrigation management (PIM). The overall purpose of WBI's program on PIM is to stimulate high-level policy dialogue on participatory irrigation management within the country, leading to policy commitment and programmatic action.

The PIM training program has thus far been initiated in India, Pakistan, and Morocco. Several other countries, including Egypt, Indonesia, Nepal, and several African countries are anticipated to join the program during 1996. Selection of countries is based on the relevance of participatory irrigation management to ongoing or planned World Bank loans and on the level of interest expressed by the host country and the concerned operational division.

For each country where it is offered, the program entails a series of three seminars (Phases 1-3) with the potential for further involvement in implementation (Phase 4) and evaluation/dissemination (Phase 5). The five phases are as follows::

Phase 1: A national seminar to introduce policy makers to the implications of PIM, consolidate national experience, learn about best practices from other countries, and formulate an indicative action plan for enhancing participation in the irrigation sector;

Phase 2: An international seminar held in a model country for PIM where participants can visit field sites to learn directly about the host country's experience and compare experiences with other participants;

Phase 3: A follow-up national seminar to formulate a national action plan for PIM;

Phase 4: Special-purpose seminars and/or training assistance to help support the implementation of a national PIM program; and

Phase 5: Support for evaluating national PIM programs and disseminating lessons learned to other countries.

Style and Methods

The seminars in phases 1-3 are designed to encourage the active participation and open discussion of all participants. Each seminar has the very practical goal of drafting an indicative or revised action plan for implementing participatory irrigation management. Presentations from international and national cases, and in particular the field visits, provide ideas to consider for including in the action plan. Much of the real work of the seminar is done in concurrent small group sessions where participants focus on a specific topic and discuss ideas in terms of their relevance to national conditions. The ideas that emerge from the small groups are presented to the plenary session for further discussion and possible incorporation into the indicative action plan.

Sponsorship and Cost-Sharing

Program activities are co-sponsored by WBI and one or more host country organizations, agencies, or ministries. The national co-sponsor normally covers all expenses for national participants (including air travel in the Part-II international seminar), with the understanding that funds can be drawn from ongoing World Bank loans. WBI generally covers the costs of international resource persons.

Participatory Irrigation Management: Benefits and Second Generation Problems

Acronyms and Definitions

Participation in irrigation management by water users can take a wide variety of forms. Farmers can be involved in various system management functions, including, planning, design, operations, maintenance, rehabilitation, resource mobilization, and conflict resolution. Moreover they can be involved in these functions at various system levels; from the field channel to the entire system.

Almost all irrigation systems have some involvement by water users in system management. When people speak of introducing "Participatory Irrigation Management" (PIM), they are thus usually referring to a change in the level, mode, or intensity of such participation that would increase farmer responsibility and authority in management processes.

Irrigation Management Transfer (IMT) is a more specialized term which refers to a process of shifting a number of basic irrigation management functions from a public agency to a private sector entity, a non-government organization (NGO), a local government, or to a local-level organization with farmers at its base. The most common form of IMT involves the shifting of management responsibility from a centralized government irrigation agency to a financially-autonomous local-level non-profit organization which is either controlled by the water users of the irrigation system or in which water users have a substantial voice in the control process.

The changes in management reported in the five primary case studies on which this paper is based can all be considered forms of IMT. However, there is an important difference between the organizational form of the recipient organization employed in the Philippines, on the one hand, and the form employed in the other four cases (Mexico, Turkey, Colombia, and Argentina). In the Philippines, the primary management unit employed is "community-based" and results from an intensive grassroots organizational campaign involving hired community organizers. This primary management unit is fairly small (less than 100 hectares), relies primarily on voluntary labour in carrying out its functions, and the most important relationships among members of the unit are social. In the other four cases, the organizational form of the irrigation systems can be termed an Irrigation Districts (ID). Irrigation Districts are typically larger (several thousand hectares), rely principally on paid employees to perform its functions, and link farmer-members together mainly through ties of economic self-interest.

Issues are analysed in this paper using a set of three basic categories:

(1)   the processes used to introduce programs of IMT;

(2)   the impacts of the introduction of a program of IMT; and

(3)   second generation problems and possible solutions.

The term "second generation" requires some explanation. Transferring substantial management authority to a locally-based organization is a complicated undertaking and may involve changes in national policy, regulations, and organizational structures; creation of new organizations at the local level; transference of equipment ownership; and changes in personnel, in addition to the shifting of management functions to the new managers. Any undertaking this complex, in addition to solving problems, will almost certainly create new problems which did not exist before or were not previously evident. An example might be inadequate technical capability of new irrigation field personnel. These problems are here termed "second generation problems." In addition, there may be situations, such as low agricultural productivity, which were present prior to the transfer, but which were not acute problems when irrigation fees were low or non-existent. For our purposes, these are also included in the category of second generation problems. Some second generation problems may be a result of faulty processes used to introduce the new management system. Some may be a result of conscious choices during implementation to defer consideration of certain potential problems in the interests of accelerating program coverage.

Others may be virtually unavoidable, though the ability to anticipate major problems in advance should allow corrective measures to be put into place earlier than would be otherwise possible. In this paper, problems are analysed from several different perspectives — that of the water user, the irrigation association, the irrigation agency, and the government. In terms of these perspectives, a change, such as increased irrigation service fees, may constitute a problem for water users, but a benefit for the irrigation association and perhaps the government, if it reduces government subsidies.

The term "impacts" also requires some explanation. In general, impacts can be either positive or negative. When they are positive, they are "benefits," or "positive benefits." When they are negative, they are similar to what are called second generation problems here. In the paper, negative impacts will be noted in the discussion of impacts, but will be discussed in more detail in the following section on second generation problems.

Process of Introducing New Forms of PIM

Background Conditions

Considering the case studies as well as PIM programs in other countries, key background conditions leading to the turnover process include:

- national budgetary crisis
- top level political will to place irrigated agriculture on a sound economic footing
- progressive deterioration of irrigation infrastructure due to deferred maintenance.

Only in one case (Colombia) of those reported at the workshop was it the farmers who initiated the process. At the irrigation systems level, another set of conditions that must be taken into consideration include:

- physical condition of hydraulic network;
- the social, political and economic conditions; and
- water availability.

Political will at the highest level of the government was a main component in Mexico's IMT program. In 1989, with a new administration in office, comprehensive water management was recognized as a top priority issue, the National Water Commission (CNA) was created and a national policy on privatization took off. In Colombia it was the National Planning Department who in 1991 submitted the Land Reclamation Program 1991-2000.

In Turkey, a budgetary crisis led to a squeeze on financial allocations to the state water resources agency (DSI) provided the initial impetus. In Argentina, a program to modernize the entire economy, which began in 1990 with the privatization of large electricity utilities, led to the turnover of water management to the provinces. In the Philippines the NIA embarked on an ambitious program in 1974 to increase rice production. A provision of this policy was for subsidies for the O&M of systems to be gradually phased out over a five year period at which time NIA would be directly dependent on collections from farmers for its O&M expenses.

The hydraulic infrastructure should be in fair condition and an affordable and reliable water supply should be available most of the time. Being in fair conditions means the hydraulic infrastructure is in operating conditions capable of delivering water to farms in sufficient amount to satisfy crop needs and in a timely manner. Surface drainage of surplus water and salinity should not be limiting factors. If these basic conditions are not satisfied, then a rehabilitation plan should be considered.

In Mexico, although the irrigation districts had suffered considerable deferred maintenance, the IMT program took off quickly because all of the systems were performing satisfactorily at the outset, with water conveyance efficiencies at main and secondary canal levels on the order of 60%.

The readiness of users to assume management responsibilities has to do with political and social factors. For example, it seems clear water users from Saldana and Coello in Colombia were "ready" in 1976 when they asked their government to turn over the administration of their districts to them. Due to its political background and social context it is apparent that Argentinean farmers were also ready for the change. In Turkey, farmers were used to a tradition of strong central government and many, especially the early adopters, were market oriented producers. This made them somewhat receptive to government declarations of the necessity of the transfer of managerial and financial responsibility.

In most districts of Mexico, however, this was not the case. In the first place there was the land tenure issue. After the 1910 Revolution land was divided between two types of small farmers, *ejidatarios*, who worked small plots of land held communally, and small landowners, some of whom belonged to the *hacendados* or landlords' elite. Second, there were numerous voracious, government controlled *ejido* leaders who, when consulted about the coming change, voted against it suspecting that the change would mean a reduction of their present status. Promotional aspects of IMT in Mexico country were vital in order to overcome these difficulties.

National Policies

National policies for IMT implementation vary both in regard to objectives and scope. Under the Philippines Water Code of 1976, the appropriation of waters by an irrigator association has priority over requests of individuals. The government then helped farmers organize themselves into irrigator associations which enter into various types of contracts with the NIA to handle O&M of the system.

In Turkey the General Directorate of State Hydraulics (DSI) is the main executive agency of the Government for the country's overall water resources planning, execution and operation. It was established in 1954 and is part of the Ministry of Energy and Natural Resources. Since the early 1960s, DSI has had a program to transfer O&M responsibility for secondary and tertiary distribution networks to IAs. Under the program IAs entered into contracts with DSI to take administrative responsibility for tasks such as collecting and submitting farmer water demand application forms to DSI, managing water distribution below the secondary canal, and cleaning and minor repair of canals and other small hydraulic structures. Although existing municipality law appears to be providing a workable initial basis for the formation of IAs, their further development and evolution may require a law specifically for IAs.

In Argentina the country is a federal entity divided into 23 provinces, which are autonomous in all aspects related to water (rights, granting, duration, taxation, etc.). IMT programs began in 1990 as a response to pressure from the national government to reduce bureaucracy and render public administration more efficient.

In Mexico the 1989 presidential decree that created CNA also granted the agency the responsibility to: (a) define the country's water policies and (b) allocate water to users through licenses and permits. The new policy:

- Created autonomous and self-financing and water utilities to provide water services in cities and in irrigation districts,
- Encouraged water reuse and water quality conservation, and
- Promoted a new water culture based on the efficient use of the resource.

In Colombia the policy of IMT is part of a larger shift in Agricultural Policy towards minimizing state subsidies and regulation. However the state continues to play a major role in land reclamation, and rehabilitation and expansion of irrigation areas.

Irrigation Service Fees

In the case of Mexico, the IMT program was part of a series of changes in the economy including reductions in subsidies for agricultural credit and inputs, elimination of guaranteed support prices for the major agricultural crops, and increases in energy and fuel prices. Transfer of O&M responsibility for the irrigation districts, leading to users paying the real cost of irrigation water, was seen as just another step in the liberalization of the economy. In a similar fashion, in the case of the Philippines, a 1974 Presidential Decree authorized the National Irrigation Agency (NIA) to delegate partial or full management of irrigation systems to duly organized associations. Under this decree, NIA was allowed to keep all irrigation service fees, with government subsidies for O&M expenses gradually phased out over 5-years. Thus, at the end of this period NIA was to be directly dependent on ISF collections from farmers for O&M expenses.

The financial aspects of IMT in Turkey are similar to Mexico in that the policy is designed to shift the burden of O&M costs from the government to the users. However, the government continues to subsidize maintenance, which is not the case in Mexico and the Philippines. Colombia has also shifted the financial burden of O&M to the users, while irrigation schemes in Argentina are under joint management with fee collection by both the government and the IAs.

Water Laws

National water laws that clearly specify the rights of the IAs and the individual users appear to be an important factor in successful management transfer. Without such rights, the IAs are extremely vulnerable to increased demands from other more powerful users, such as industry and municipalities.

Beginning in 1976, the Philippines, attempted to develop a water rights register of all water rights in the country, including specifying in volumetric form the water rights of the national and communal irrigation systems. As part of the registration process all IAs must register their water rights. Once legally registered, water rights cannot be withdrawn except for failure to use them as stipulated in the law.

This system can be contrasted to that of Mexico where the IAs are given a limited concession to use the irrigation infrastructure and the associated water supply, but do not have a clearly specified legal right to a volumetric supply. With domestic supply having priority, the IAs are not ensured a constant water supply over time. The concessions are also for a fixed time frame (20-30 years) after which they can theoretically be reassigned to another user. Since none of the concessions have expired to date, it is unclear exactly what process will be used to determine granting of a second concession.

The country where IAs appear to be most vulnerable to changing water demands is Turkey, where water is controlled by DSI and the individual associations have no water rights. This system works in areas where there is little competition for water but leaves the associations extremely vulnerable in areas where municipal & industrial use is expanding rapidly. Colombia is somewhere in between these two situations but still is highly dependent upon the national irrigation agency, as the country has yet to establish a legal water rights system.

The Mode of Implementing PIM

The process of implementing a change to a more participatory type management system varies widely from the bottom-up approach used in the Philippines to the more top-down approach used in Turkey and Mexico.

In promoting the PIM process, Colombia has invested much less time and effort than the other countries. As a consequence, the process of transfer was very quick but there have been second generation problems. Users have been less well informed and have been uncertain about their rights with respect to ownership and changes in management practices. As a number of the irrigation systems in Colombia are based on river pumping, some of the more recent cases of transfer have involved additional negotiations with respect to energy subsidies.

Argentina has used information meetings and word-of-mouth to make a rapid transition to PIM, while Mexico and Turkey have used more structured informational meetings with users. These two countries also invested heavily in training their own staff. In Mexico, audio-visual materials prepared by IMTA and outside firms were used to persuade users that IMT was a positive change.

The Philippines has used the slowest process with intensive use of institutional development officers and farmer organizers (FOs) to serve as catalysts. These organizers lived in the villages and organized exchanges between NIA and the users. However, NIA has now realized that too much dependence upon FOs has slowed the process and has now reduced significantly the number of FOs.

In terms of transferring responsibility, most of the Colombia IAs appear to have had very little say in the process. However, this has changed somewhat as more recently the IAs have been negotiating energy subsidies and insisting upon rehabilitation prior to transfer.

In the Philippines there has been more dialogue through the organizers but given that only limited transfer of responsibility has taken place, this approach has not achieved rapid transfer of responsibility to the users. In contrast, both in Turkey and in Mexico there have been very active negotiations concerning transfer terms. As a result the IAs have been able to exert some power and develop a partnership with the agency based on a meeting of equals. In a number of cases they have also been able to ensure that critical investments were made prior to transfer.

As a key stage of the transfer process, Mexico, Colombia and Turkey have adopted a phase of shared management between the agency and the IAs during the transition. The duration of this shared management phase varies by country and by district but is usually between 6-12 months. In contrast, in Argentina the systems were transferred much more quickly, while in the Philippines there has been a very gradual shift in responsibility, in some cases over a period of many years.

Type/Nature of IA

The type and nature of the IAs are directly related to the structure of the economy as well as the type of irrigation found in the countries. In Mexico, Japan, Taiwan, Turkey, Argentina and Colombia, the IAs are larger (2,000-50,000 ha) and are organized more along the lines of commercial entities, reflecting the more commercial nature of the irrigated sector. Agriculture is developed on a cash basis and many of the staff are hired professionals paid in cash from the ISF. Given their large size, the IAs can afford to purchase and maintain their own transport and maintenance equipment.

In contrast, IAs in the Philippines are very small (100-300 ha) and are often organized based on the village structure. Most of the labour is voluntary labour provided by the users, and very few, if any, of the irrigation staff are hired professionals. Given the small size of the IAs there are diseconomies of scale and hence the organizations cannot afford to own specialized maintenance equipment. In the Philippines, irrigation service fees are usually paid in grain and therefore are very awkward to store and transport and typically result in 10-15% losses due to damage during storage and transport.

In Argentina, IAs are public NGOs with full legal authority, including the power to tax. The IAs in the other countries have a limited power to establish and collect ISFs but do not have any other local taxation powers and therefore the majority of their income is from ISFs.

Impacts and Benefits of Transfer

Implementing a program of management transfer is a complicated undertaking which involves incurring costs and affecting the lives and livelihoods of many people. It is thus not desirable to enter into such a program unless the benefits of the changes are positive and significant. Impacts, of course, may be either positive or negative, and they can be either qualitative or quantitative. And because the change in management patterns will usually occur simultaneously with other changes in physical, economic, and social conditions, it may be difficult to separate the effects caused by management changes from those caused by other factors.

The nature of the impacts which occur will be shaped by the social, political, and economic characteristics of the countries involved. Impacts are also conditioned by the perspective from which they are viewed.

Important perspectives in this case include those of water users, the associations they have already created, the irrigation agency and the national or state government under whose overall control these systems operate. What is positive from one perspective might be negative for another. Judgment is thus required in evaluating the overall impact of a program, and the tradeoffs in positive and negative benefits among the various groups affected.

Legal and Organizational Impacts

Some of the countries first build on a new national water Law before moving into a transfer program (Mexico, the Philippines and

Colombia). Others move into transfer supported only by old national laws (Turkey) or local laws (Argentina). There has been a lot of discussion on the issue of water rights from the perspective of ownership versus concession. In most of the countries, all water belongs to the governments, either at federal or provincial levels. Irrigation Districts were usually created based on constitutional mandates that clearly defined the system's physical limits and thus the water right implicit to it. In some countries such as Argentina, however, the right of individual appropriation takes legal precedence.

After turnover, irrigation associations (IAs) in some countries receive the concessions which give them access to collective water rights. This does not change the existing system of water rights and priorities. In other countries, systems of water rights allocation and protection are virtually non-existent, which can lead to serious second-generation problems. In most of the case study countries, agricultural uses are subordinate to municipal uses, which include domestic and industrial use.

Sense of Ownership

The formation of IAs generally creates a stronger sense of ownership of the system on the part of users. In Japan, the Philippines, Colombia, and Argentina (Mendoza) this has been the case for many years. In Mexico and Turkey this change has taken place more recently and has been an exhilarating sense of pride for some users. Even though actual ownership of system facilities remains with the government.

In Colombia some consideration is being given to transferring actual facilities ownership to IAs, an idea supported by some INAT staff and IA officers. This idea has also been discussed in the Philippines. There are a number of problems with this step, including loss of the right of eminent domain which is held by governments only and allows land condemnation for system expansion, and increased exposure to liability for accident and injury on the part of the IAs. Such a move may also weaken an IA's claim on public assistance for rehabilitation and emergency repairs.

A sense of ownership is also enhanced because users now have a greater voice in selecting the governance of their system, normally by electing boards of directors through the IA general assembly. With leaders elected in a democratic manner, needs can be expressed more readily. The opposite situation is presently occurring in Taiwan, where the users are turning systems back to the government. This is a result of the declining importance of agriculture in the Taiwanese economy and the high costs of system O&M which cannot be sustained by many IAs.

Transparency

All of the countries report that their transferred irrigation districts are now managed with more transparency, meaning by this that major decisions are exposed to the association's members or their representatives through board of directors and general assembly meetings. In Mexico, a representative assembly, formed by delegates selected by the general assembly takes routine decisions without summoning the general assembly, which may be composed of hundreds or even thousands of users. In Turkey a similar body made up of 30 to 70 local government officials and farmers also serves as a general assembly.

Transparency may be even more important in financial accounting than in organizational politics. Before transfer, the government collected fees and was generally supposed to invest the fees back into the system where they had been collected. However, due to bureaucratic procedures collected fees were not necessarily applied in the same systems where they originated. This resulted in complaints by users, and the agency was seldom forthcoming with information on financial accounts relative to particular systems. With associations collecting and managing funds, and with accounting done on the basis of the unit managed by the IA, users can get a better understanding of how their money is being used. However, strict supervision and auditing of water fee collection and expenditures by an outside party is still necessary to counter the possibility of money being diverted for unauthorized purposes.

Where the transferred units do not include an entire hydrologic unit, it may be more difficult for users to develop a sense of the disposition of the fees collected from them due to sub-division of responsibilities among a number of IAs. In this case, accounts should separate out the portion of fee collections that are earmarked for O&M within the IA unit, and the portion allocated to higher level O&M. An alternative might be for the IAs to merge into a larger unit. However there a number of other factors to consider before advocating such a step.

Water Fees Increase

Usually one of the immediate effects of IMT is an increase in water fees at the irrigation district level. In some countries the increase can be very dramatic, on the order of 100% to 200% or more (Mexico, Turkey). However to make a valid comparison, fee rates must be put in constant value terms before being compared. Irrigation service costs are generally considered appropriate if they constitute 5% to 8% of total production costs. In most countries, even after transfer, they tend to fall into this range. In the case of small holding sizes and lower value crops, however, the gross margin retained by farm households is also a relevant factor, and where this is already small, a doubling or tripling of ISF can create hardships for small farmers and their households.

A proposed increment in ISF may also have a negative psychological impact on water users that have come to expect heavily subsidized irrigation service. This kind of attitude may be gradually changed with promotion activities before transfer takes place and, of course, with efficient financial management and better service once the association is formed.

ISF collection rates are above 50% in all five countries currently implementing IMT activities, and are above 70% in four of the five. However, in looking at the share of O&M costs covered by ISF collections, the picture is a bit different. For example, before transfer in Mexico, the ISF collection rate was around 100%, but this amount covered only 25% of O&M costs. After transfer, the ISF collection rate is still 100% in transferred districts, but fees now cover close to 90% of O&M expenses. In order for transferred schemes to be financially self-reliant, they must achieve high rates of fee collection as well as having an ISF that covers O&M costs.

Subsidies

From the government's perspective, an increase in O&M cost recovery normally means a reduction of subsidies. By subsidy we define here any kind of costs associated with the provision of irrigation services that the farmers should be paying instead of the government. Expenditures within an ID are usually related to:

- O&M costs plus overheads,
- acquisition of premises,
- purchasing machinery and equipment,
- rehabilitation and/or modernization,
- technical assistance and training, and
- emergencies.

O&M plus overheads refer to expenses related to the daily administration of the system from the head works down through the tertiaries. These expenses have to be paid in full, either by the government, the users, or a combination of both if a system is to be sustainable. The sharing of costs between the government and IAs ranges from 40/60 to nearly 0/100, as is the case with Mexico.

Acquisition of premises refers to the land and buildings IAs use as they provide service to farmers. In Mexico CNA initially provided premises to IAs as part of the concessions. After six years of turnover though, some associations have purchased their own offices, which they often refer to with pride.

Providing machinery and equipment to IAs has been a key component of transfer programs in Colombia and Mexico. In other countries, like Turkey and the Philippines, machines are still under the control of the irrigation agency. In Mexico equipment provision was heavily subsidized by the government to encourage farmers to participate in the IMT program. Many of the machines given in concession to IAs were old and the government had to subsidize repairs. Other machines were purchased under a World Bank loan. As lowering of maintenance costs was a concern for both government and users, emphasis was put on the purchase of light machines to substitute for the heavy construction equipment which had traditionally been used in the ID by CNA.

Rehabilitation and/or modernization of hydraulic infrastructure is also a key issue for all countries involved in IMT programs. Although farmers usually ask for investments of this kind to be made before turnover takes place, governments are often unwilling to supply the necessary resources to fulfil these requests. Turkey is an exception to this rule, where rehabilitation was not an integral part of the IMT program. In some cases countries use international loan funds for this purpose, enabling the government to negotiate with farmers the amount of needed investment and/or the financial share that users will need to contribute. Costs are often shared between the government and IAs on a 50/50 basis (e.g., Mexico).

Support for technical assistance and training may also be subsidized through IMT programs. In Mexico, technical training for both IAs and government irrigation district staff, has been provided by CNA with assistance from the Mexican Institute for Water Technology (IMTA). More recently, starting in 1995, some IAs have signed private agreements with IMTA in which technical assistance is being provided on specific projects with a limited subsidy from the CNA. In the Philippines, promotional and technical support to farmers is regularly provided by the government through various types of community organizers. In Turkey, extensive training for agency field staff and farmer leaders was provided by the irrigation agency. Yet, in all countries the training needs for PIM programs by far exceeds current investments in training.

Subsidies are also normally available in the case of emergencies situations. Although little specific information on this is available, all of the countries recognize the need of governmental intervention when such cases do arise, e.g., hurricanes in Mexico or typhoons in the Philippines. Water scarcity also creates emergency situations for transferred IDs. For example, during the 1995-96 irrigation season, an extraordinary drought took place in northern Mexico. Due to the lack of water, IAs did not have water to deliver to farmers; consequently, and collection of fees dropped nearly to zero. For many of the IAs the government had to step in with 100% subsidy programs to avoid a collapse of the associations. A few IAs that were only partially affected by the drought tackled the problem by hastily imposing a compensation fee on their farmers to implement a water reuse program.

Impacts on Operational Procedures

It is expected that a shift to participatory local management will improve the effectiveness and efficiency of water delivery as well as the quality of maintenance. The case studies provided some details to document these impacts but there is still a need to obtain better information about the operational impacts of IMT.

Improved Maintenance Programs

One of the immediate consequences of IMT programs is that irrigation systems have improved maintenance programs. (Colombia, as recently reported, seems to be an exception). A recent study carried out in 1994 in Mexico by the Colegio de Post-Graduados reported that 84% of the sample users believed maintenance had improved substantially since IAs have been in charge of O&M. In some areas not only have IAs completely eliminated the problem of deferred maintenance, but are also are now putting surplus money from O&M activities into programs for the modernization of hydraulic infrastructure.

Improved Services

Although few specific studies on quality of services after IMT exist, the general impression is that, after turnover, services have substantially improved in regard to timeliness, reliability, and equity of water distribution. There are many reasons to believe this. In the first place, after turnover users usually have better access to the irrigation service provider. Compared with the relationship with the agency that existed before, the farmer-association relationship is much better in terms of distance, personal contact and feedback on complaints. Water conflicts among farmers also tend to be minimized because farmers usually search for solutions to their problems at the IA level which is physically closer to them. There is also a degree of self-control prevalent, as the users know each other very well and are therefore able to regulate behaviour among themselves.

In a 1994 study of IAs in Mexico, 84% of the users indicated that water distribution had improved since turnover, 79% said they were receiving enough water, 79% were also receiving water in time, and 64% indicated water was being allocated in the appropriate amount. However, water conflicts do arise between the IDs and IAs, especially when water is scarce. Recent lessons from the Mexican drought indicate that, under extreme competition for water, IAs were overwhelmed by operational problems. This required prompt government intervention to keep the conflicts from getting out of control and becoming an explosive political issue.

Agency Re-organization

IMT programs can have a significant impact on agency organization. A sharp reduction in field personnel and O&M staff usually takes place. A change in the role of the agency with regard to water management in the systems follows.

Staffing: Once the government's employees perceive the possibility of losing their jobs as a consequence of IMT implementation, they may attempt to block the turnover process. Clear legal arrangements between the parties are needed in order to overcome difficulties. Unfortunately, as the number of people involved in the process may be quite large. In Mexico, 5,000 out of 7,000 government personnel were to be released from the IDs because of turnover, while the Philippines has reduced government agency staff from 19,353 to 10,368.

Changing the role of the government irrigation agency is a natural outgrowth of IMT. The change requires not only willingness from the part of high-level officials but also training for them on specific issues. One is simply convincing governments officials that there have been substantial changes in the way the irrigation sector is structures. For instance, in some countries the agency keeps asking from IAs the same types of information as if management had never been transferred. In Mexico, after six years of turnover, changes in the governmental sector are still incomplete. Once the turnover thoroughly takes place in a system, the agency's new role should be more oriented towards supervision, guidance, monitoring and regulation of water rights, plus selected technical assistance support to IAs. Management responsibilities of the system should be in the hands of the users, with the possible exception of head works which when considered of strategic importance or technically sophisticated, should remain under agency's direct control and administration.

Second Generation Problems and Solutions

As noted in the introduction, problems and solutions in a particular location will differ, depending on the perspective of the observer. The discussion below addresses problems from the perspectives of (a) the irrigation association, (b) farmers, (c) the irrigation agency, and (d) the government. Solving these problems can involve a variety of steps including revision of laws or implementing regulations and changes in organizational structure, organizational rules and processes, and funding mechanisms.

In addition, associations will require supporting services. Support services are services which come from outside the association itself but which are necessary for it to carry out its mission. They include such things as financing for equipment purchases, legal advice, computer programming assistance, and financial auditing services. Such services may be difficult for an association to generate internally (e.g., financing for heavy equipment) or be used only infrequently, (e.g., specialized maintenance equipment) and hence be too expensive for an association to maintain on a full-time basis.

Support services can be procured by associations from a variety of sources—private firms, public agencies, universities and institutes, non-government organizations, and regional or national federations of associations. In the past, it has usually been assumed that any such services must come from government agencies. Today it is recognized that higher quality and less expensive services may often be obtained from other sources, and that the government should generally serve as only one of several alternatives, rather than the sole source.

Irrigation Associations

Insecurity of water rights was identified as the most serious second generation problem affecting irrigation associations in the five case study countries. Water rights which are often absent, poorly-defined, or insecure, can (a) inhibit investment in new system facilities or rehabilitation, (b) encourage short-term thinking and behaviour on the part of association managers and farmers, (c) result in heavy expenditures in legal costs to defend a poorly-defined water right, and, ultimately, (d) lead to a reduction in water supply and even system collapse.

An effective water right should provide security to the association, but must, at the same time, be adaptable so that water can be diverted to other more productive or higher priority uses as economic and demographic conditions change. In this event, there must be a provision for appropriate compensation to those who are giving up the water right by those who gain it. In Mexico, for example, an association's irrigation water right is always subservient to present and future municipal demands. This creates considerable insecurity for associations which share water sources with growing municipalities and violates both the principles of security and that of just compensation.

An effective water right should be specified in both quantitative and qualitative terms. Water quality degradation by upstream effluent discharges, as from a factory or an inadequate municipal sewage treatment plant, can render the water unusable by downstream irrigators. It can also make the water suitable only for lower value crops, since biologically or chemically contaminated water may not be permitted for production of higher value fresh fruits and vegetables. This will become an increasingly serious problem as water reuse increases in response to growing demand from all sectors.

Establishing a water rights system where it is lacking, as in Turkey, or clarifying water rights where they are weak, ineffective, or inequitable, as in Mexico, will usually require action from the national legislative body or from top level executive leadership, or both. It is thus extremely important for water user associations to have adequate representation of their interests when these issues are taken up.

Two different types of support services are identified as being crucial for associations in attempting to establish or clarify their water rights. The first of these is legal advice and representation when the association faces challenges to its rights. Such representation is best secured from private law firms, if available, since services secured from government sources may be of lower quality, and may be subject to political pressures which would compromise their objectivity.

Legal representation is also important during the formative stage of an association, when negotiations with the government irrigation agency will establish the contract or concession which will control the relationship of the association and the government. Unfortunately, associations which are just forming may be unaware of the importance of high quality legal advice at this stage or may be unable to afford it. A national federation of associations can play a valuable role as a source of legal advice and assistance to newly forming associations.

The other type of support service required by associations is lobbying on their behalf in government policy making councils. Since other interests, such as municipalities and industrial water users, are usually larger than individual associations and likely to be more powerful politically, it is important for associations to establish regional or national federations representing many associations and a large number of farmers. This will give them political influence with which to counter the lower of competing interests. A federation would also be to represent irrigation associations on the board of directors of the national irrigation agency as is currently the case in Colombia.

Financial shortfalls comprise another high priority second generation problem affecting associations. A central feature of the PIM programs undertaken in all of the case study countries is financial autonomy. Financial autonomy is the condition where an organization generates all of the revenue it needs to support itself and to perform it's primary functions.

It implies that the association is not directly subsidized by the government, or that if it is subsidized, that the subsidy is a fixed amount which does not vary according to the condition of the association's balance sheet. The principal source of revenue for most associations is irrigation service fee (ISF) collections. Financial shortfalls affect a number of associations in Colombia and Argentina have been a concern in Mexico since the 1995 economic crisis.

Financial shortfalls are a function of several factors, including ISF rates, ISF collection effectiveness, the contribution of other sources of revenue, and expenditure patterns. One important factor is the structure of the ISF. Fees can be levied on flat rate or volumetric bases. A recommended structure for fees is a two-part one comprising both fixed and volumetric components. The flat portion would constitute a "connection charge", a charge for simply being within the boundaries of the system's service area whether or not water was actually taken from the system. This would reimburse the association for expenses incurred in maintaining the physical and administrative capacity to deliver water to the farm. The absence of this component in the fee structure of Mexican associations has created severe problems during years when drought greatly reduced the available water supply to the system's water users. The other portion of the fee would be based on the volume of water actually delivered during a cropping season, or some proxy for this amount, such as area irrigated and number of irrigations given. This would cover the costs incurred by the association which are related to the amount of water given, and would serve to limit excessive demand for water.

Revenue from ISF is also dependent on the percentage of the fees assessed which are actually collected, though associations in many countries do a reasonably good job in this regard. An exception is the Philippines where low collection rates have been a persistent problem for NIA.

Solving problems of revenue shortfalls that relate to fee levels and collection efficiency is largely an internal association responsibility. Outside assistance may be useful, in some cases, in estimating farmers ability to pay particular ISF levels, and in analysing management systems set up for collecting revenue. This is discussed further in a following section.

Underlying difficulties in generating sufficient ISF revenue to sustain system operations, in many cases, is the low productivity of irrigated agriculture in system command areas. Low productivity can result from a large number of factors, but is often associated with small farm size, a subsistence orientation, production of low value crops such as grains, inappropriate agricultural policies, a poor natural resource base, and inadequate agricultural support services.

In such cases, a solution to the association's financial problems may be possible only if the underlying problems in the agricultural sector are addressed. If these problems cannot be solved, then the options are for the government (a) to have other agencies provide technical assistance to increase production, or (b) provide the association with special subsidies. As a last resort, the government may have to consider taking back the responsibility of system management and financing. However, as irrigation service fees typically constitute only 3 to 10 percent of total production costs, reducing them will generally not solve underlying problems of high agricultural production costs and low productivity.

Rehabilitation is identified as a third important second-generation problem for irrigation associations. All irrigation systems require periodic rehabilitation and modernization. While usually less expensive, in real terms, than the original construction, rehabilitation is a costly undertaking, and is usually beyond the financial and technical means of an association to undertake. Seldom is there a clear and consistent government policy on responsibility for rehabilitation. In the absence of such a policy, the tendency is for associations to defer needed rehabilitation in the hope that the government will step in and take responsibility for it. In this case, IAs usually underinvest in system improvements between rehabilitations. This tendency is reinforced by the government retaining ownership of system physical facilities, while transferring to associations only the use rights of the facilities. Associations may thus regard the responsibility to rehabilitate those facilities as belonging to the government unless a different policy is clearly stated in the agreement between the government and the association.

A related problem is that of cost sharing between the government and the association for rehabilitation. Because irrigated agriculture benefits people beyond the ranks of system irrigators, and because full coverage of rehabilitation costs is usually beyond the means of the irrigators themselves, a sharing of costs is appropriate. Responsibility for even a share of the costs involved will tend to counteract the tendency of an association to defer maintenance, as noted above.

To cover its share of future rehabilitation costs, associations usually need to accumulate a capital replacement fund over a number of years and there needs to be a legal basis for establishing such a fund.

There should also be incentives for establishing and contributing to such a fund. Unfortunately, fiscal and monetary policies in many countries such as Turkey and Mexico, have led to high rates of inflation and low or negative real interest rates on savings, which acts as a powerful deterrent to fund accumulation.

Governments may need to create special investment opportunities for associations which allow them to earn reasonable rates of return on accumulated funds. Likewise, there should be incentives for associations to make improvements in physical infrastructure. One way to do this is to establish a trust fund, perhaps with donor financing, from which associations could request funds to complement their own investment funds.

The matching ratio for such a funding facility should be established and made known in advance. A number of support services are required specifically to support system rehabilitation. These include (1) assessments of system facilities, (2) credit, and (3) design and construction assistance. Regular assessments of the condition of system facilities can be done jointly by the association and the government agency, as in Turkey, or may be contracted out to an engineering consulting firm acceptable to both the association and the agency. Such assessments can be used as a basis for annual maintenance planning, to suggest the need for selective improvements in system facilities, and for planning whole-system rehabilitation.

If an association is unable to accumulate its share of rehabilitation cost prior to rehabilitation and does not have the ability to assess a special charge on the members, it will need a source of credit. Credit can come from private banks, government banks or other lending facilities, or from insurance pools in the case of rehabilitation induced by natural disasters such as floods, typhoons, or volcanic eruptions. Such a credit facility could also be used as a source of financing for capital equipment needed for system maintenance.

Rehabilitation will also require external technical services for design and construction. Because of the sharing of costs, both the association and the government should be involved in decisionmaking relating to the selection of consultants and contractors and monitoring their performance. Advice and guidance to the association on handing these tasks might usefully be given by a federation of associations, since rehabilitation occurs only infrequently in any one association.

Lack of financial and administrative management expertise is the final topic identified as a high-priority second generation problem by workshop participants. There are several possible responses to this problem. One would be skill enhancement through staff training programs. Skills can also be enhanced by replacing less skilled people with more capable ones. Contracting out for specialized services is another important way of addressing management deficiencies in associations.

One extremely important step in improving the quality of association management is to increase the transparency of management processes. This has a number of positive effects. It can (a) reduce the potential for misappropriation of funds, (b) help insure that salary levels and benefits are realistic, (c) help insure that maintenance allocations are appropriately targeted, (d) reduce favouritism in making personnel appointments, and (e) improve responsiveness of association staff to users. A number of steps can be taken to increase transparency in association management. These include:

- regular external audits of financial accounts,
- use of standardized budgeting and accounting frameworks,
- wide dissemination of simplified budgets, plans, and financial statements,
- active involvement of the board of directors in forward planning, budgeting, and auditing, and
- broad representation from among users on the association board of directors.

There is broad scope for employing external support services to improve management of the association. Services that may be required include:

- advice on establishing and revising management systems and procedures,
- advice on establishing financial budgeting and accounting systems, including software,
- establishment of standard budgeting and accounting formats,
- standards and requirements for regular external audits, and
- management training.

These services can be obtained by the association from a variety of sources, including private firms, a national or regional federation of associations, NGOs, government agencies, and universities and training institutes. The government is the logical party to establish mandatory standards and requirements for external audits, but the audits themselves, could be done by a private firm of chartered accountants. Other services, such as management and accounting system advice could come from a variety of sources, with private sources being generally preferred.

One argument in favour of the provision of these services by government agencies will often be that they can be obtained at no or low cost. What makes this lower cost provision possible, however, will generally be implicit government subsidies to the service providers. A preferred alternative would be to provide the funds supporting these subsidies instead to the associations as grants to be used for obtaining management support services. This would allow the associations to contract for these services among alternative providers.

The demand-driven competition thus induced would be a very healthy force acting to hold down service prices and improve quality of services delivered. Provision of such grants during the transition phase from government agency to association management could be a very useful institution-strengthening activity.

Irrigation Agencies

Dislocation of staff is the most prominent problem experienced by agencies following irrigation management transfer to users. This problem is typically dealt with in several ways.

First, O&M staff levels are reduced by attrition. When positions become vacant due to retirement or resignation, they are left vacant or filled by internal transfers rather than new hiring.

Second, financial incentives are often provided for early retirement of older staff. Third, existing staff are transferred to other positions which become vacant rather than filling them from outside the agency. In some countries such as China, where it is difficult to lay off staff, sideline enterprises have been created which can generate income for the irrigation district and cover the salary costs of the involved personnel. In the Philippines, this has taken the form of an irrigation consulting company, NIA Consult, which was created as a subsidiary of the National Irrigation Administration to provide irrigation advisory and design services within the Philippines and abroad.

In some cases, redundant agency staff may also be re-employed by the newly created associations which take up management responsibility for the schemes. Such employment should be at the discretion of the association. Loss of technical capacity in the national irrigation agency is a common problem faced by agencies which transfer management significant irrigated areas to associations and experience loss of technical staff. To address this problem, agencies can

- obtain specialized expertise from outside consulting firms as needed,
- increase salaries to attract and retain high-quality staff,
- provide in-service training opportunities for staff, and
- revise job descriptions to bring in new staff with the desired qualifications.

Defining a new role for the agency is another important challenge. With their operational responsibilities transferred to associations, the agencies need to design a new role to address emerging problems. Doing this requires discussion among staff within the agency and also at higher levels of the government, with broad participation by all involved parties. The aim should be to build broad consensus and political commitment for the new role. In some cases, changes in legislation may be required to enable the assumption of new responsibilities.

The new mandate should contain a clear definition of roles and responsibilities and should define skill requirements to carry out the new responsibilities. It should also contain a timetable for accomplishing the shift to the new mandate. Support services which could be useful in this process include:

- comprehensive diagnosis of the agency/association relationship and the associations' support needs,
- professional assistance with the agency's strategic planning process, and
- consulting services to design new management information systems for the agency.

In Colombia, the national agency, INAT, is employing professional consultants to help them define a new role for themselves under an Inter-American Development Bank credit.

Farmers

Second generation problems experienced by individual farmers relate mainly to the need to increase farm productivity to pay higher irrigation fees and to take advantage of possible improvements in irrigation service quality. Support services required may include:

- production credit,
- extension advice,
- new technologies,
- markets and market information,
- access to inputs, and
- post-harvest services.

Although government agencies are the traditional source of many of these services, in many countries, private or other organizations are playing an expanding role in supplying some or all of the services listed above. There is also the question of the potential role of the association itself in providing other agricultural services, in addition to irrigation service. As a general rule, the association should demonstrate competence in its core activity of irrigation management before considering such ancillary activities as providing other agricultural inputs.

Government

The principal second-generation problem for government, beyond those already identified for the irrigation agency, is the reduced control which it will have over irrigation activities at the system level, and a diminished ability to use irrigation as a tool to implement other national policies and priorities. An example might be the government's wish to promote cultivation of upland crops rather than rice during a particular season.

In the past it could work through the national irrigation agency to adjust water delivery schedules and volumes to try to achieve this end. Following transfer, this becomes more difficult. There are other tools, such as support prices and subsidies, to achieve the same ends, however, so that losing control of irrigation system management should not pose a significant problem for agricultural policymakers.

Summary and Conclusions

Experience is now available from a number of developing countries which have recently implemented Participatory Irrigation Management (PIM) programs and there is additional evidence from developed countries which transferred irrigation management functions to locally-based organizations many years ago. In the case of the developing countries, while the overall benefits of PIM have been positive, in some cases second generation problems have manifested themselves and, consequently, have tended to reduce the magnitude of the potential benefits. In the process of introducing PIM programs, political will at the highest level is a critical background condition for a rapid and sustained transfer program. A second important element is that the irrigation infrastructure be in fair condition so that it could deliver irrigation water as required. A suitable legal framework is also necessary for the sustainable functioning of the transferred systems. Lack of clear water rights has led to second generation problems including conflicts with municipal and industrial users as well as other irrigation organizations.

PIM is designed to shift the financial burden for irrigation service from the agency to the users. This aspect has to be made very clear when the process of transfer is introduced. Failure to address the financial side of system management is a primary cause of second generation problems. In general, the countries that have a clear policy on irrigation service fee rates and collection practices have sustainable water user associations.

The type and nature of the associations are very dependent on the structure of the broader economy as well as the type of irrigation and the tradition of management prevailing in the country. Where economies are more developed and diversified and irrigation systems are large, associations have tended to evolve successfully. These associations are generally large and can hire their own staff and own specialized irrigation equipment. In contrast, in countries where economies are less developed, agriculture is more subsistence oriented, and irrigation systems are small, the associations tend to be small and more problematic in terms of management.

These changes in management structures and processes have had important impacts both positive and negative on four important target groups: farmers, the irrigation association, the irrigation agency, and the government. For example, while increased service fees have reduced the financial burden on the government and increase the sustainability of the IAs, they have added to the costs of production for the users. On the other hand, from the perspective of the farmers, transfer has resulted in a sense of ownership, reduction in conflict and improved maintenance. Transfer has reduced the O&M staff of the irrigation agencies, and consequently the overall number of civil servants working in the irrigated agriculture sector. However, in a number of cases, this has also reduced governments' control over cropping patterns and over water resources more generally. This reduced government control has generally led to greater farmer satisfaction, more transparency in decisionmaking, and greater overall economic efficiency.

Changes in management responsibility have led to second generation problems in most countries, some of which are already affecting the involved parties while others loom as potential future problems. Insecurity of water rights was identified as the most serious second generation problem affecting transferred systems. The primary solution is to establish a secure legal right that has both quantitative and qualitative dimensions. Federations of associations can provide assistance and legal representation.

Financial shortfalls comprise another second generation problem. The principle source of revenue for the association is the irrigation service fee. A two-part fee consisting of a fixed connection charge and a volumetric charge can provide greater financial stability. Increasing the productivity of irrigated agriculture is also an important element in improving the financial health of associations. Outside assistance may be useful in analysing and improving management systems set up to assess and collect service fees.

Rehabilitation poses a number of second generation problems. In the absence of a clear and consistent policy on rehabilitation, maintenance is often deferred. There is a valid argument for developing a cost sharing formula where the government pays a share and the users pay the remaining share. If possible, the IAs should have a sinking fund for accumulating funds to cover their share of future rehabilitation costs. When this is insufficient, it is important to have an outside source of affordable credit. Other needed supporting services include assistance with maintenance assessment practices and technical design and construction services. Effective financial and administrative management of the associations requires specialized staff training and increased transparency. Support services such as external audits, and standardized accounting packages can also contribute to better management of the associations.

Irrigation agencies also suffer from a number of second generation problems. Dislocation of staff, loss of technical capability, and the need to define a new role for the agency are all problems found in countries that are instituting PIM programs. In particular, the problem of what to do with excess staff is a difficulty faced by almost all agencies. Solutions include attrition, retirement incentives, creation of specialized consulting units, retraining and assistance with job placement, and transfer to other units. Along with the problem of staff displacement, agencies also face the problem of the loss of specialized skills. These may be replaced by outside private consultants but may also require the agency to provide specialized training for existing staff.

Second generation problems of farmers are related to the need to increase agricultural productivity, including the need to shift to higher value crops. Services such as credit, agricultural extension, market access, technical inputs and post-harvest assistance are all needed. In some cases these services will come from the government but increasingly from the private sector as well. Federations of IAs can also play an important role in the provision of such services.

A shift from public agency control to local participatory management is unlikely to happen without some second generation problems. Rapid institutional change will almost always require corrective measures to address unexpected problems. Countries with flexible policies and procedures will be able to address these problems, as they arise. This report has summarized a number of solutions countries have employed to address second generation problems associated with the shift to participatory irrigation management.

3

Population and Irrigation Water Management

In order to develop successfully, a country must take account of global and regional conditions and pursue a development policy that combines economic growth, sustainable use of natural resources, and improvement of the quality of life of its people. In Morocco, an agricultural-pastoral country, where water is scarce and half the population lives in rural areas, the interaction between humans and nature is particularly intense, for three reasons:

- For farming families, which make up 80 percent of the rural population, the place of residence is also the workplace, and the separation between life in the home and in the production site is not clearly marked, as it is in the city.

- Rural areas are directly exposed to constraints imposed by climatic variables and water supply, especially with respect to seasonal activities, income contingencies, and spatial mobility.

- The wide geographic dispersal of homes, encouraged by the customs of the country, with its inevitable negative consequences for basic infrastructure and various community installations, contributes to the socioeconomic and cultural isolation of families, and is an obstacle to the overall development of the country. Families are forced to lead a more traditional life that is less open to modernity.

Moreover, it should be emphasized that the rural family is caught in a kind of vicious circle, based on four elements:

- the need for free family labour;
- demographic pressure on natural resources;

- poor purchasing power; and
- poor profitability of investments.

As a result of the small scale of economic enterprise and low rate of income in rural areas, women and children are required to work extensively and without pay, putting upward pressure on birth rates as a source of labour. However, demographic growth has in turn created great pressures on forests, roadways, cultivated lands, the economic viability of farms, and so forth, and consequently on opportunities for investment and increases in productivity and revenues. Therefore purchasing power remains poor, as does access to social services such as education, health, and leisure, making it even more difficult to increase economic activities and improve production and revenues.

Since Morocco became independent more than 40 years ago, these demographic characteristics have led authorities to give great importance to population policy and management of natural resources, particularly water.

This chapter will have three parts: demographic policy (past record and future outlook); water policy (the management challenge); and a case study of the management of irrigation water in a small-scale hydraulics area.

Population Policy: Past Record and Future Outlook

Morocco has taken a census of the population every 11 or 12 years since 1960. It took more than 50 years, from 1900 to 1952, for the population to double, from five million to 9.3 million. The growth rate, a mere 0.6 percent at the start of the century, steadily increased to 2.8 percent between 1952 and 1960, before declining to a level of 2.06 percent between 1982 and 1994. This recent decrease is explained by:

- an increased rate of use of modern methods of contraception, from about 20 percent in 1982 to 50 percent in 1995;
- a relatively greater decline in fertility than mortality (the synthetic fertility index was 5.52 in 1982 and 3.52 in 1994);
- an increase in the percentage of unmarried women between 20 and 24 from 40 percent to 60 percent; and
- an increase in emigration.

Although the average population density is 36.6 persons per square kilometer (km^2), there are great disparities between regions, on the one hand, and between cities and rural areas, on the other. In fact, in the northwest region, which includes the coastal Tangiers-

Casablanca axis and the interior cities of Fez and Meknes, one sees that less than three percent (2.56 percent) of the area contains almost a third of the population. Moreover, the growing frequency of drought years since 1980 (nine out of 16 years) has intensified the exodus from country to city (270,000 persons per year between 1988 and 1992). Thus, the growth rate for urban areas has reached 3.61 percent in spite of low fertility (2.56 percent), while the rural growth rate has not surpassed 0.67 percent in spite of a high fertility rate of 4.25 percent. The result of this migratory movement is that the proportion of urban dwellers has gone from 42.7 percent of the population in 1982 to nearly 51.4 percent in 1994.

The developments described above are the results of a population policy, followed since independence in 1955, that is characterized by the abrogation in 1955 of a law against the use of contraception, the liberalization of abortion (1967), the revision of the legal status of women, and the implementation of several Population Education activities dealing with such issues as spacing of births, improving literacy, and health and hygiene.

Demographic Outlook

These educational efforts have produced a number of important results, especially in the area of urban demographics. However, demographers expect the population to continue to increase at a rapid rate, mainly because the population is still young with a consequently strong reproductive potential. In fact, the percentage of young people under the age of 15 has only slightly declined, from 41.4 percent in 1982 to 37 percent in 1994. Projections for the year 2014 show that the overall population will reach nearly 35 million, an increase of one third from today. The rural population will remain at current levels, while the urban population will increase by 70 percent to account for about two thirds of the total population. Long-term estimates show that the population will double towards the middle of the next century, reaching about 50 million, posing major challenges in the areas of nutrition, satisfaction of social and economic needs, and the rational and sustainable management of natural resources. Morocco is fully aware of the challenges implicit in these trends.

Water Policies: Managing a Challenge

Morocco intends to build on its past policy achievements and redouble its efforts within the new global context, in accordance with prevailing concepts of sustainable human development.

Water and Rural Development

The concept of rural development necessarily implies reduction of current disparities between urban and rural areas. Reduction of rural socioeconomic deficiencies is directly related to sustainable water management, whether for irrigation or for domestic and industrial consumption. Necessary measures include:

- extension of roadway networks, electrification, and telecommunications;
- improvement of rural habitats, domestic installations, potable water supplies, and sanitation;
- job creation and anti-poverty campaigns in small rural centers;
- improvement of sanitation and hygiene;
- expansion of family planning and spacing of births; and
- improvement of rates of schooling and other efforts against illiteracy.

Implementation of these programs is inconceivable without universal and scientific water management.

Spatial and Temporal Variations in Water Availability

Water management in Morocco, like everywhere else, is tied to the management of other natural resources, and must address the needs of its three major users: agriculture, industry, and the household sector. With two maritime borders (the Atlantic Ocean to the west and the Mediterranean Sea to the north), Morocco is relatively well supplied compared to other countries of North Africa. In addition, the mountain ranges, which cover a substantial part of the national territory, act as reservoirs.

Annual rainfall is estimated at some 150 billion cubic meters (m3) overall. However, two constraints must be noted: rainfall variation in time and space. Morocco has always had drought years, but their frequency and severity have greatly increased since the early 1980s. Of the last 16 years, nine droughts have been recorded, whereas during the first half of the century there was on average only one drought every ten years. Spatial distribution of rainfall in Morocco is characterized by declining gradients from north to south and from west to east. Certain regions receive 600 to 700 millimetres (mm) per year, while others receive less than 100 mm.

Three basins on the Atlantic (Sebu, Bouregreg, and Oum Rbii) hold two thirds of the freshwater potential. Waters flowing towards the Mediterranean or towards the Sahara nearly disappear at times.

Water Distribution by Sector

Climate and land contour determine both the state of vegetation and the natural resources management policy followed by the government. The total area of the country, 71 million hectares, is divided into usable agricultural land (13 percent), forest and alpha zone (12.5 percent), roadway terrain (30 percent), and uncultivated land (44.5 percent). As a result of management practices, 80 percent of the 150 billion m^3 of precipitation is lost each year through evaporation or discharge into the sea. Only 14 percent (21 billion m) of the total rainfall is presently believed to be usable by acceptable economic and technical means. Currently 11.7 billion m^3 are used, of which 75 percent is surface water and 25 percent is subterranean water. This water is divided as follows:

- Agriculture: 86 percent, permitting year-round irrigation of about 900,000 of a potential 1.35 million hectares. To this surface water is added some 430,000 hectares worth of seasonal water and flood water.

- Industry: 5.5 percent, especially concentrated in the northwest zone of the country and in the Casablanca-Mohammedia agglomeration.

- Potable water: 8.5 percent, i.e., an annual average of 38 m^3 per capita, or over 100 litres per day, for each of the 26 million inhabitants.

The Water Challenge

The overall per capita water supply figure, based on the total amount of potentially usable water, is 800 m^3 per year, placing Morocco in the poor category (500 to 2,000 m^3 per person per year) in international terms. The figure for actual use (all uses) is 460 m^3 per year, for a weak international rating (100 to 500 m^3).

Water supply is therefore a great challenge for a country whose population is going to increase by one third by the year 2014, and double towards the middle of the next century, with increasing urbanization and industrialization rates. Additional factors making this challenge all the more serious include the following:

- Soil erosion and degradation of ground vegetation reduce the soils water-retention capacity;

- Farmers are still not rapidly adopting techniques and equipment that economize water irrigation;

- Water pollution from agricultural, industrial, and household activities is increasing, while strategies to combat the problem are only beginning to be developed; and
- Water management to date has been dominated by government agencies, and the necessary involvement of civil society (the general population, professional organizations, and selected non-governmental bodies) will take time to organize.

Case Study of Irrigation Management

In accordance with the International Monetary Funds Structural Adjustment Policy, the Moroccan government is attempting to reduce its involvement in economic activities and to entrust them to the population concerned. The example that follows illustrates some of the difficulties involved in applying this policy on site. The project in question involves the state granting farmers control over management of a small-scale water works area. After a brief review of popular participation and the reasons for promoting the management scheme in Morocco, we will deal successively with a description of the project zone, its contents, the possibilities for development and revenue creation, and, finally, a critical analysis of the project in its present phase.

Popular Participation in Irrigation Management

A reading of Morocco history clearly shows that its people have always tried to manage its water supply well in order to survive times of drought and famine. Traditional collectives were formed long ago for water management purposes, and have continued to this day, in spite of modernization efforts undertaken by administrations during the French protectorate years and after independence.

The survival of this management method can be explained by the ingenuity of the works construction, the respect given to the precepts of Islam, local customs, the principle of water rights for each irrigator in the collective, and the operational rules of the management group (Jmaa or Dioune), based on justice, equity, voluntary service, and penalties for delays.

Each ethnic faction chooses its naib (delegate) to manage its seguia (canal). The Amazal (water men) and Maujari (assistants) are paid in kind or in water allowance. At the beginning of the Protectorate, in 1914, a water commonality law was passed. This law was followed in 1924 by another law creating favoured Agricultural Syndicate Associations (ASAPE), adapted to the needs of the colonials, i.e., assertion of administrative control and the ability to expropriate, institution of water fees, implementation of voting rights according to property size, etc. After independence in 1958, a general law was passed to encourage creation of associations for the promotion of popular participation in all areas of development. This law coincided with the implementation of major irrigation programs, and the government took the opportunity to create its own irrigation associations.

Results did not meet expectations, and in 1990 the government passed a new and more specific law, creating Agricultural Water-Users Associations (AUEA). These associations were intended to take charge of irrigation works created by the government. As had been the case in the past, the government took the initiative, defined the rules of the game in its own way, and maintained the right of oversight of the associations operations.

The most recently passed law (1995) introduced new options, such as water basins and the anti-pollution campaign. It confirms the AUEA law, but does not allow its effective implementation. Everything leads us to believe that Morocco is still looking for the best way to induce popular participation. The 1995 water law established the principles of vested interests and public property rights. On the institutional side, the law retains traditional structures, but does not designate a specific supervising ministry, allowing for decentralization and introducing basin agencies using government workers. However, despite some indications of progress, the new water law is deficient in a number of areas, especially the following:

- water pollution;
- water and environment protection;
- water and landed property; and
- regulation of certain spacesuit does not address subterranean water, deep subterranean water, or coastal zones.

Although the law introduced the principle of the polluter pays, the question of the price of water is still at issue, and questions relating to setting its price have not been resolved. The decrees outlining the laws application, now in preparation, may serve to remove these deficiencies.

We are still facing the difficult questions of how to optimize use of water resources and guarantee social justice in the future. Growing scarcity and the decline in water quality clearly challenge the viability of traditional technical solutions and show that a global, multi-disciplinary approach is urgently needed.

Equally clear is the need to institutionalize principles based on the concept of integrated water resources management, i.e., management incorporating social, political, economic, legislative, and other institutional actions to guarantee the optimal quantity and quality of this resource for all users. These actions may take the form of institutional coordination, multi-disciplinary research, human resource training, information dissemination, popular participation, development of appropriate technology, using and improving local know-how, and engaging in intersectoral or international cooperation.

Integrated management seems to have become the popular catch phrase with regard to protection of the environment. There is no doubt that effective protection means consideration of the total ecosystem and its interactions, not just consideration of the different elements in isolation. Efficient and effective management of water resources therefore requires that we not neglect any of the interests dependent upon it, nor any of the physical laws that govern its workings.

Development of Small-Scale Hydraulics in Morocco

In Morocco there are two types of irrigation: Large-scale hydraulics (GH), involving vast areas fed by high-capacity dams and providing year-round water supply (presently about 500,000 out of a potential 830,000 hectares); and small-scale hydraulics (PMH), involving small areas of several hundred hectares fed by water sources that are not highly regulated (e.g., pumps, water diversion, co-lineal reservoirs, spring water catchments, and flood waters).

Government interest in PMH dates to the 1960s, but increased in the mid-1980s because of frequent droughts, and enabled by credits granted by the World Bank, the German Kreditanstalt fur Wiederaufbau, and state subsidies. Estimates show that there are about 813,000 hectares of PMH areas, including:

- 383,000 hectares irrigated by year-round water (47 percent);
- 265,000 hectares irrigated by seasonal water (33 percent); and
- 165,000 hectares irrigated by flood waters (20 percent).

The goal of the state is to reduce the amount irrigated by seasonal waters to 170,000 hectares, and increase the amount irrigated by year-round water to 510,000 hectares (60 percent). This measure should contribute in a major way to nutritional security, job creation, and the effort to slow rural exodus throughout the country. The goal will be reached through rehabilitation and modernization of equipment in the areas concerned, using traditional irrigation systems based on customary rules of water distribution.

These steps are generally accompanied by socioeconomic measures, such as canalization of the potable water supply, electrification, improvement of the public information system, intensification of land ownership, and substitution of more profitable crops for those that are now widely cultivated. Greater organization and participation of the population are highly desirable additional components, supported by the passage of several new laws.

Has the intended goal been reached? The case study that follows attempts to contribute to answering that question.

The Project Zone

Demographic data, climatic problems, and socioeconomic conditions of the project zone are more or less similar to those described above for rural areas in general. We will therefore limit our description here to additional specific elements pertaining to the site of the project, the state of property structures, and agricultural development. These elements are indispensable in evaluating the degree of success or failure of the project.

Site of the Project

Morocco possesses eight major water basins, one of which, Sebu Ouengha-Beht, in the Gharb region in the northwest, is a currently regulated resource supplying about two billion of a potential five billion cubic meters of water. The area presently encompasses a year-round irrigated surface of 106,500 hectares, including 90,000 hectares irrigated by GH and 16,500 hectares irrigated by PMH. The basin potential will make it possible eventually to irrigate nearly 250,000 hectares 220,000 by GH and 27,500 by PMH. The PMH zone chosen for the case study is located in the foothills of the Rif mountains, about 60 kilometers north of the cities of Fez and Meknes, south of Sidi Kacem, at the confluence of the Western Inaouen and Middle Sebu rivers. Measured rainfall in this area is between 400 and 600 mm per year. The project requires supplying 15,000 hectares with two trenches; the first 6,500 hectares includes areas known as Sectors II and III, and the second 8,500 hectares includes Sectors I, IV, and V. We will focus exclusively on Sector II (2,700 hectares), since the work there is relatively advanced.

State of Property Structures

There are 920 distinct properties in Sector II, at a very low average of about three hectares each. The total number of owners, some of whom also own land outside the sector, is estimated to be 1,800, or an average of two owners per property.

Fifty percent of the properties are less than one hectare, representing in all no more than six percent of the total area. Properties of less than five hectares are excused by law from direct participation in equipment investment. Properties in this category represent 88 percent of all Sector II properties and 31.5 percent of the total area. Trench properties, five to 20 hectares in size, represent 9.5 percent of all properties and 25.5 percent of the area. Two percent of the properties are more than 20 hectares each and amount to 37.5 percent of the area. Many of the properties, 43 percent, covering 41 percent of the area, have a legal status of shared ownership, and therefore lack a well-defined spokesman. When the General Assembly of Associations met, only 60 percent of the owners attended; in other words, the number of absentee or non-resident owners is large.

Current State of Development

Forty-four percent of the properties are irrigated in whole or in part, i.e., irrigation is nothing new for the area. The sector is served by a series of small hydro-agricultural units, called oulja, fitted into the bends of the river, offering a certain technical and socio-economic homogeneity. There are a total of nine such units: five on the right bank and four on the left bank.

The lands are currently cultivated using a very low-grade technology. Some farmers own tractors and do the soil work for themselves and others, but this work is often slow and of poor quality; fertilizers, if they are used at all, are used without soil analysis, and plant sanitation treatments are mostly nonexistent. The major crop is still cereal (60 percent tender wheat); edible legumes are only 15 to 30 percent of the total. Livestock raising is predominantly limited to sheep, as this practice makes use of straw, which is the fallow plant in drought years, and sometimes barley fodder. Some owners have a few (six or seven) milk cows of local or mixed breed.

These cultivation practices are explained by the climate and the low level of technical expertise. Farmers say they have one good year out of every three. Yet the zones potential is far from negligible. Some farmers dig wells or pump directly from the river in order to have year-round water. Others create additional irrigation channels. As a result one finds rich crops on certain properties, including mint, market gardens, fruit trees (olives and citrus), and alfalfa.

Some farmers obtain record yields of tender wheat, up to 5,000 kilograms per hectare. However, we should point out that they manage their water so as to use as little as possible, only using reserves when the climate begins to harm the crops. According to the farmers, the amount of water used does not exceed 6,000 to 7,000 m^3 per hectare per year. This type of management requires great flexibility in decision making and farming operations, and the risks are not insignificant.

The Project

The projects objective is systematic development of the area through modernization of the irrigation system, including the improvement of its management by establishing user associations specifically for that purpose. The technical study carried out in 1984 called for building principal installations (including five pumping stations, eight recovery stations, and delivery pipes) and open, gravity-driven irrigation networks, internal and external drainage systems, and electrification. Much of this work is now 80 to 85 percent complete.

Investment costs for the sector are estimated at 210 million durham (DH), equivalent to about US$40 million. The average cost per hectare is therefore on the order of 80,000 DH (US$16,000). Direct user participation accounts for 30 percent of investment costs, or 24,000 DH (slightly less than US$5,000). At a lending rate of four percent over 17 years, the yearly payment amounts to 2,050 DH (US$410). As noted above, farmers owning less than five hectares are exempt from this obligation, and those owning up to 20 hectares are exempt for their first five hectares.

Creation of the AUEA and their Union

To ensure takeover of the installation operations by the local population, it was decided to establish four Agricultural Water Users Associations (AUEA) for the nine oulja. The hydraulic equipment includes an upstream portion common to the four associations, and a downstream component proprietary to each one.

The upstream portion includes pumping and recovery stations, transfer canals, regulating reservoirs, drainage networks, and the delivery paths to the works. The downstream components include networks for irrigation and drainage, and paths serving the plots of each association. Thus there are two levels of system management. The responsibility of each association is relatively clear with regard to their own proprietary downstream components; however, association responsibilities are not so clear for the common, upstream elements.

In order to address this problem, two plans are being considered:

- Creation of a non-public company whose task it will be to deliver water to the associations. This plan compensates for the lack of experience on the part of the local population in the management of sophisticated and expensive collective installations. On the other hand, it reduces the involvement of the associations, which is contrary to the spirit of the law. The law establishing the associations requires that they be fully responsible for operation of the equipment they use.

- Creation of a union of the four associations, which would be able to enter into contracts with third parties for certain services.

Whichever plan is adopted, oversight will be necessary on two levels: internal association oversight, particularly for water payment, whose uniform price will come from a cost-sharing arrangement among all users; and an external oversight, by the state, for the protection and maintenance of its property.

If the "Union Plan" is adopted, it should provide for:

- A governing body, composed of a general assembly of all users and a council made up of the presidents and a few delegates from all the associations, plus a government representative;
- Internal regulations establishing the rights and obligations of the associations and the union with regard to financial and technical management and types of control; and
- A contract between the state and the union specifying the conditions of the installations grant.

Costs and Fees

Operational costs must be calculated in order to determine the billing price of water for users, and to determine the advantages they will derive compared to their current practices. We will not go into detail, but will only consider the broad issues separating the Union and Association categories.

Operational costs include a technical team and an administrative team for the associations and the union. Equipment costs include vehicles as well as office and computer material. Energy expenses are calculated on the basis of 10,000 m³ of water per hectare per year (lower than the 14,400 m³ figure used in the original study), which corresponds to a four-year rotation system (50 percent fodder, 21 percent cereals, 25 percent market garden), requiring 100 m³ of water and an irrigation efficiency coefficient of 84 percent.

Anticipated expenses for periodic maintenance, major repairs, and replacement of equipment have been estimated on the basis of established norms. The annual cost is estimated at 13.5 million DH

(US$2.7 million)-2.9 million DH for the associations and 10.6 million DH for the union. Considering the preceding data, we can calculate the cost, which is divided into a fixed fee and an energy fee:

- A fixed fee, indexed to exploitation charges, estimated at 4,370 DH (US$874) per hectare, of which 1,195 DH (US$239) is for the associations operations and 3,175 DH (US$635) for the unions operations.

- An energy fee, tied to each stations consumption, estimated at 1,220 DH (US$244) per hectare. For reasons of simplicity, the same amount will be charged to all users in the four associations.

Therefore the user will have to pay annually per hectare the following amounts:

- 1,195 DH (US$239) for the association's operations;

- 3,175 DH (US$635) for the union's operations, collected by the associations;

- 1,220 DH (US$244) for energy, to be paid to the union;

- Subtotal: 5,590 DH (US$1,118) per hectare for water, about 0.6 DH (12 US cents) per m^3; plus

- 2,050 DH (US$410) for direct participation in investments, payable directly to the state;

- Total: 7,640 DH (US$1,528) per hectare.

Future Development

Based on research results and current practices within the region, projected yields have been set for each crop, as have charges for each crop, excluding irrigation. The margins determined, excluding irrigation, show a disparity of as much as a factor of seven or more between high-margin crops (e.g., mint and garlic, at 60,000 DH, or US$12,000, per hectare) and low-margin crops (e.g., wheat and fodder, at 7,000 to 9,000 DH, or US$1,400 to $1,800 per hectare).

This calculation allows us to conclude that the greatest economic hope does not lie in improving low-margin crop yields, but rather in adopting rotations based on high-margin crops, and raising the intensity rate of cultivation. The combination of these two variables, using the calculation of margins per rotated hectare, allows us to determine the crop system that will best exploit the water charges (55,000 DH, or US$11,000 per hectare). We observe therefore that the more crop intensity increases (110 to 130 percent), the more the cost of water in relation to charges and revenues declines (40.6 to 28.9 percent and 26.7 to 14.6 percent, respectively), and the more the margin produced increases (34.2 to 49.5 percent).

Will the project improve the current intensity rate, which now varies from 80 to 100 percent, as well as increase the income of farmers and their management ability? The analysis that follows will help us form a more exact response.

Analysis, Suggestions and Questions

The data cited above lead us to certain analyses in our effort to acknowledge and surmount some of the obstacles to realizing the projects goals. This chapter does not hope to find precise solutions to all the problems; we will be satisfied to offer some helpful suggestions and to pose some critical questions in order to highlight certain lines of thought.

The very modern concept of hydro-agricultural equipment, with its complexity and expense, marks a dramatic departure from the current practices of the population and from current socioeconomic conditions. Notwithstanding the importance of a reliable water supply, will the population accept high annual costs for participation in investments that were decided without their consent, water fees based on consumption rates, and unit prices that exceed those that are customary to the region? Moreover, it is unfortunate that certain measures that were supposed to accompany the hydraulic installations reallocating land, adapting and energizing local technical administration units, involving the private sector were incompletely realized, if at all. It could also be asked if the sectors burden could be relieved by including lands owned by users outside the sector.

The general principle of state disengagement should not be applied without consideration of local realities. Further subsidies should perhaps be provided to ease the burden on users, and payments deferred until the AUEAs are firmly established and land development has begun. A broad campaign to educate and persuade farmers is indispensable. As for land development, it is clear that progress has been made in the areas of productivity and rate of intensity, as well as in marketing of products. Aside from the need to reflect on the compatibility of the farmers free choice of crops to grow and the existence of a single irrigation system, the problem of the technical framework remains unsolved. What will be the states role; the private sectors; the AUEAs?

While we regret that the steps taken in these areas since the projects beginning in 1983 have been very modest, we wonder if the solution does not lie in the current phase of the state-private sector relationship, i.e., the former contributing infrastructure and personnel, the latter capital and energy. In any case, the solution for the future lies in the adoption of the activity by the population itself, within the framework of the AUEAs or similar organizations.

Land improvement will also require substantial investments at the production level. The National Agricultural Credit Bank will need to adapt to serve this category of farmer, for whom the narrow and complex property laws are not very favourable. In addition, these cultivators should be the first to benefit from the Agricultural Development Fund, which is intended to support production. Let us point out that the user organization is an inevitable prospect, since otherwise the population cannot participate in development. Without popular participation, sustainable development cannot be assured.

Allow us to point out as well that organizational initiatives rarely come from the population under the socioeconomic conditions that exist in rural areas. The government is forced to take the initiative, hoping the population will follow. The law on AUEAs reflects this dilemma. On the one hand, there is the governments duty to initiate and maintain basic installations, and on the other hand there is the governments desire to transfer management, within an organized and democratic framework, to a local population that, unfortunately, is not ready to handle it.

Thus there arise questions and criticisms regarding the law governing AUEAs and the conditions of its application. The creation of irrigation works preceded the creation of the AUEAs; consequently, the latter cannot be held responsible for the works technical deficiencies and poor adaptation to local circumstances.

The AUEAs, as government creations, are imposed on the people and do not reflect their own choices. The government is an official member of the elected association council. Is it necessary for the law to require all users to be summoned to the general assembly, and not to be able to have valid deliberations unless two thirds of the owners are present or those present own half the total surface of the area? Is it even possible? Is it realistic and functional to insist on the principle of one user one vote with such a heterogeneous population?

Who can one interview when a property is in a legally confused situation (lease-holders, partnerships, absentee owners, joint tenancies, rights of tenure without true ownership)? Is it appropriate to give the association the ability to punish users, even with expropriation?

In reality the problem is one of creating the best fit between the local manpower, the characteristics of the population, and the contents of the law they are required to respect. This is why we raise the question of whether this project would be better conceived within a broader developmental program in which other elements will be considered, such as family planning, literary programs, childrens education, emancipation of women, job creation, and the campaign against subdivision of the land.

The government must be able to act in an effective, dynamic, and intelligent manner in order to exploit and protect the investments it has made, and to win the populations confidence and gain its adherence, through such means as intensified training and information programs, mobilization of influential local leaders, and reduction of population pressures on land resources.

The situation urgently calls for intervention, but the task is complex and requires perseverance.

Conclusion

Since mid-century, Morocco has seen a fairly rapid increase in population that has not been accompanied by sufficient socioeconomic development. The deficiency has been especially marked in rural areas, with direct repercussions on the management of natural resources. Water, which has been vital in sustaining increased food production, threatens to become scarce if demographic growth continues its rapid pace.

Therefore the challenge for the future is to maintain the water-population balance at a level that does not impede the countries development. The population must contain its growth rate and raise its socioeconomic level, and every possible technique must be developed to save water and protect it from pollution.

This goal cannot be realized without the genuine participation of the population alongside the government. Since the problem is rather complex, all the agents of development must contribute in a complementary, coherent, and dynamic manner, and with the requisite perseverance. Given the problems of communication between the government and the farmers, there is reason to believe that the solution lies in the involvement of a non-governmental organization, acting as an interface between the two parties, and serving as a coordinator for the involvement of other agents. This possibility deserves careful study.

Paddy Irrigation and Water Management in Southeast Asia

The evaluation covers six gravity irrigation schemes with reservoirs for water storage in Myanmar, Thailand, and Vietnam. Widely dispersed in the region, the schemes were chosen for their variety, rather than their typicality. Though paddy predominates in all, four of the schemes are large—at least 40,000 ha—and the other two are small.

Four have plenty of water; the other two, in the central dry zone of Myanmar, have much less than planned. To compare the organization and effectiveness of operation and maintenance (O&M) between irrigation and flood control, the study also reviewed the performance audit findings from flood control schemes at three sites in the evaluators visited farmers and officials at the scheme sites, and public irrigation authorities responsible for the schemes. Interactive group and household interviews were arranged in all four countries. The evaluation focused on agroeconomic impacts, and on operation and maintenance issues.

Agroeconomic Impacts

In the Irrigation Schemes

Crop yields have been close to expectations.

- In four schemes, including three of the large ones, irrigation covers much smaller areas than planned. The main reasons are planners' over optimism, engineering errors, lower-than-normal rainfall, and failure to extend the tertiary canals.
- Cropping intensities are much lower than expected at three sites and are falling at a fourth.
- Only one scheme—the small Azin scheme in southern Myanmar—has reached its targets for both area and cropping intensity.
- In two thirds of the schemes, output is much less than expected when the projects were appraised. Output of paddy, and of a few other major crops at the two schemes where paddy is not completely dominant, is from 32 to 73 percent of appraisal estimates in five schemes. Again the exception is Azin in Myanmar.
- The collapse in international rice prices since the early 1980s, after the projects were appraised, helped drive down

profitability. In real terms, the price of rice in 1995 was only one-third of the 1995 price that was projected in 1980. Farmers could have coped more easily with the price decline had they diversified their cropping pattern, as planned in four of the schemes. Instead, they grew more paddy.

* Economic rates of return are no higher than 7 percent in all schemes and are negative in one.

By the Bank's standards, guided by opportunity costs, these have been uneconomic investments. But borrowers point to the visible signs of substantial intensification of agriculture and increases in yields over large areas that were previously rainfed—as well as the considerable indirect regional and social benefits of the investments that are not captured by rate-of-return estimates. They are committed to maintaining the schemes.

Welfare Impacts

In most cases, incomes from paddy in the project areas are well below appraisal estimates. In Vietnam and Thailand, actual incomes are only 10–30 percent of appraisal estimates. The gap is lower in Myanmar, but mainly because appraisal projections were less ambitious.

Farm incomes in the project areas may not be high enough to keep families, and especially their youth, committed to farming. Net household incomes from irrigated cropping on average-size farms in the six schemes range from $600 to $2,000 a year, depending on the size of the farm, local market prices for paddy, and the extent of diversification out of paddy. Unless these farms adopt higher-value cropping systems, their sustainability is in doubt. The relation between low paddy prices and rising wages suggests an irreversible trend. In Thailand, the world's largest rice exporter, economic modernization is pulling farmers into factories even as low paddy incomes push them out. Vietnam and Myanmar, also rice exporters, can be expected to follow the same route.

Operation and Maintenance

Through field surveys, the evaluators assessed the performance of agencies and farmers in operating and maintaining the schemes. They observed the condition of canals and control structures, agency activity in allocating, distributing, and maintaining the flow of water, and the strengths and weaknesses of O&M by farmers. In the irrigation schemes, they found that:

- Water appears to be generally well managed. In the three major schemes, water-use efficiency ratings are high, at 43–52 percent. This compares with the 25–40 percent ratings that the International Food Policy Research Institute quotes as typical in Asian irrigation schemes.

- Though errors were made in design engineering, there is no evidence at the six sites that original construction was so poor as to make maintenance difficult.

- Agencies generally operate and maintain irrigation structures quite well, at least as regards operational plans. Dams, main canals, and structures on these canals under agency responsibility are in good order, except that many measurement devices have been removed or allowed to deteriorate. In Vietnam, the knowledge and hands-on involvement of field engineers from the two provincial irrigation services at Dau Tieng; in Thailand, the purposive reforms promoted by the Royal Irrigation Department's new project manager at Lam Pao, backed up by technical assistance; and in Myanmar, the intelligent management of scarce supplies by the Irrigation Department at Kinda, are impressive.

- Sophisticated measurement and allocation programs promoted by consultants have been abandoned, both in Myanmar and Thailand. They are premature for paddy cultivation, given the loose institutional arrangements in place and the high water tolerance of rice.

- Assessed against engineering design, irrigators behave poorly. Assessed against their collective self-interest on individual watercourses, their behaviour makes sense. In all the schemes surveyed, farmers maintain their canals to serviceable standards. Vandalism and neglect affect mostly structures that are ill suited to community needs, such as tertiary gates that interfere with the flexible operating protocols favoured by farmers, inlets that induce excess flooding in the lower reaches, and embankments that prevent drainage. When siltation and weed infestation threaten irrigation, farmers readily mobilize labour to clean up watercourses. They handle routine maintenance and minor repair of tertiary canals, channels, and associated structures on a collective basis with a modicum of support. The evidence confirms that farmers will act collectively in the common interest if substantial benefits, broadly available to the irrigator community, are at stake.

- Farmers share water. Crop yields differ little from the heads to the tails of watercourses. Farmers' own customary arrangements allow significant advantages to those at the heads of canals, but provide enough water for those at the tails. Even where water is short, relations between farmers at the heads and tails of canals are quite civil and accommodating, suggesting that enough "social capital" is available to overcome collective action dilemmas.

Role of Water user Groups

Water user Groups (WUGs) are not Functioning as well as Expected

Nominally, water user groups exist throughout the command areas of the six irrigation schemes, and are responsible for O&M below the turnouts of tertiary canals. Public agencies retain ultimate responsibility for operating and maintaining the tertiary gates, but usually share these jobs—especially the operation of gates—formally or informally with unfederated tertiary water user groups. The water user groups vary in type and effectiveness. In the internationally assisted sections of the scheme at Lam Pao, Thailand, both the water user groups and the federated groups of WUGs organized along some of the distributaries clearly show the improvements in the irrigation system, and on farms, that can follow effective organization. None of the other schemes attains this level of performance. Elsewhere at Lam Pao, the WUGs accomplish their basic purposes—to keep the tertiary canals and watercourses open and to assemble labour to help the agency keep the larger canals clear—but do little else. In Myanmar, the WUGs are subordinate to the village councils and do not seek or achieve any higher purpose. In Vietnam they are barely more than arms of the provincial irrigation authority.

Strong water user groups are not a primary cause of the relatively successful O&M observed in the schemes studied. Groups with broad participation and strong leadership enhance the efficiency of water distribution and use. But weak groups do not condemn schemes to utter inefficiency. Farmers cooperate to achieve at least basic O&M goals regardless of the maturity of the formal organization.

The experience shows there are other viable organizational models for allocating water than formal irrigator groups. The field studies identified a wide range of organizational procedures, including some reasonably well administered systems developed locally, or with targeted TA from donors, that combine hierarchical authority with user participation and are shaped by country traditions. At Dau Tieng and the left main canal at Kinda, for example, public agencies ration water intelligently according to availability, and farmers cooperate, inside or outside formal associations, to keep channels open.

Relationships between tertiary units are more problematic than those within. At this level, neighbourhood cohesion, and the other social forces that ensure reasonable cooperation within each tertiary system, are too weak to guarantee equitable water sharing. Here associations and formal federations of primary WUGs can make a substantial difference. (At Lam Pao, as the associations of WUGs sharing the same secondary canals gain strength, the functions and prominence of the watercourse WUGs themselves tend to diminish. This is predictable, because once the association of WUG leaders has determined an appropriate water-sharing formula, or cleaning schedule, meetings at the lower level can be dispensed with.)

Contrast with Flood Control

The flood control schemes in Bangladesh present a different profile, more in keeping with the paradigm of poor maintenance and lack of cooperation among farmers. Actual benefits are closer to expectations in these schemes than in the irrigation projects, but maintenance standards are deplorable and the sustainability of the structures and benefits is in doubt. Professional incentives at the Bangladesh Water Development Board favour civil engineering skills over water management skills, and the involvement of agriculture and rural development agencies is minimal. The board has only recently made efforts to organize farmer groups for maintenance, and farmers have not associated spontaneously. Though they would all benefit from proper maintenance of older schemes, the benefits of cooperation are less certain, being reaped only when floods appear, and they are not equally shared. The potential implications for the economic and social sustainability of the major flood control investments taking place in Bangladesh are disturbing.

Looking to the Future

The turning of the terms of trade against paddy growers in the early 1980s casts a shadow over plans for better O&M. At Lam Pao, farmers who let their land lie fallow in the dry season, instead of double cropping, do not join the working groups. In Thailand and increasingly in Vietnam, families who continue to farm must contend with husbands and young adults leaving the fields and wives, and older members taking up the slack.

Such trends imply a shift back to subsistence production, and less interest in or ability to do good operation and maintenance. Dry season production will be the most seriously affected by these changes, but the monsoon labour profile is also changing and with it attitudes to operation and maintenance.

Conclusions and Suggestions

Though the sample is small, the similarity of the findings across the different schemes suggests that the following lessons may have wider application:

- Tailor the prescriptions of programs for improving O&M. Agencies and irrigators do well at some functions and fail at others, often a reflection of the incentives they face. For example, exhortations to farmers to keep tertiary gates in working order, and thus risk curtailing their own water supplies, are unlikely to succeed. Hence the need to identify poorly performing components, provide incentives to bring them to appropriate standards, and tailor prescriptions based on intensive consultation with farmers and officials.

- Simplify technology. Sophisticated water distribution and monitoring technologies should be put aside in favour of controls that need less human intervention, at least until intensive diversified cropping systems are in place.

- In Bank projects, emphasize capacity building for effective water distribution associations, giving priority to federating user groups beyond the tertiary watercourse level. Hybrid organizational arrangements that take careful account of existing social networks, and that combine community labour with official agency support, should be piloted to improve the maintenance of canals and gates. Flood control and drainage schemes, where the development of user groups has been ignored, are prime candidates.

- Ensure that project engineering takes adequate account of hydrological, topographical, and social factors. If farmers are to take over responsibility for financing and developing tertiary networks, or managing irrigation, they should be involved early. Even if not formally organized, they should be brought into the design process of the irrigation system and then persuaded to enter agreements for partial financing, approval of designs, participation in construction, and management after construction is completed. Participatory project design is important everywhere but should be mandatory in flood-prone, poorly drained, densely populated areas.

- In government policy, favour crop diversification and intensification, supported by enhanced extension and marketing services. Exhortations to recover costs from farmers should be muted until water systems are reliable, more remunerative crops are introduced, and volumetric water delivery becomes practicable.

Sustaining India's Irrigation Infrastructure

Water, as an input to agriculture, is critical for sustaining the food security. India faces the daunting task of increasing its food grain production by over 50 per cent in the next two decades. Increasing competition for water in agriculture, industry, domestic and environment uses is already manifested in inter-and intra-sector, basin, state, district and village level conflicts.

These will escalate further as India's annual per capita water availability goes below water scarce threshold level of 1700 cubic meter within the next two decades. In six of the country's 20 major river basins (with less than 1000 cubic meter of annual per capita availability), water resources are under stress and depleting. By the year 2025, five more basins will become water scarce and by 2050, only three basins in India will remain water sufficient. Supply expansion, to meet expanding needs, is constrained by availability and rising economic and environmental costs associated with its development and use. The status of irrigation infrastructure and prospects for its sustainability, both physical and financial, for future water-food security is the issue under focus.

Irrigation Infrastructure

Existing Status

Since 1950, India has made direct public investment of Rs. 88100 crore in providing major, medium and minor irrigation infrastructure with an irrigation potential of 91 MHa. India Water Vision, 2025 estimated the gross water demand for multiple uses to double in 25 years from now with corresponding investment needs of Rs. 20000 crore per year. As of now, India's irrigation infrastructure is expanding by 1.8 Mha of irrigation potential with a public outlay of Rs. 7000 crore per annum. Current annual expansion is one-third less than the maximum growth achieved in the past. Deceleration in irrigation potential created through major and medium schemes started during

1980s as a consequence of declining real government expenditure on this sector.

Amidst competition from non-agricultural uses in households, industry and environment, supply of irrigation will have to keep pace with the targeted annual agricultural growth rate of over 4% in the Tenth Five Year Plan. To achieve this growth rate, irrigation sector should grow by at least 5% per annum, given 1% growth in rainfed sector, Demand-supply management in water sector and efficiency in its every use is critical for providing sustainable water-food security to the country.

More importantly, existing and expanding irrigation infrastructure has to be physically and financially sustained for improving their efficiency. Yet concerns are emerging on the physical condition of the irrigation infrastructure created so far.

Vicious Cycle

India's irrigation sector is caught in a vicious cycle. Inadequate funding for O&M over years has resulted in the neglect of maintenance and upkeep of the irrigation system leading to deterioration in the quality of irrigation service. Physically, the irrigation and drainage system is not able to receive and deliver the planned quantity of water matching with the demand pattern. Poor irrigation service, often not matching with the crop water requirements over space and time, results in low productivity of crops and income to the irrigators. Resultant dissatisfaction coupled with weak institutional linkage leads to under assessment of demand for water rates as well as low recovery of whatever is assessed. Progressive fall in the cost recovery increases revenue deficit causing adverse impact on O&M funding for maintenance works.

Deferred maintenance of surface irrigation infrastructure over years has led to further deterioration of its physical service. This is witnessed by stagnating or falling irrigation coverage affecting agricultural growth in several regions. Surely, with future expansion in food production growth critically depending on the performance of irrigation sector, what is happening to the physical status of existing and expanding irrigation infrastructure does not augur well for India's future food security and agriculture performance.

Canal Irrigation System

Despite annual expansion in the created potential and capital investments in irrigation sector, the area irrigated by the major, medium and minor irrigation systems has been either stagnating or declining from mid-1980s or during 1990s, as witnessed in some major states.

For instance, currently in Uttar Pradesh canals are irrigating 30.6 lakh ha in TE 2000 as against 33.3 lakh ha in TE 1985[6.] Similarly, in Andhra Pradesh, canals now irrigate 11% less area than what was irrigated 15 years back. Bihar, Orissa and Tamil Nadu also recorded similar decline in the canal irrigated area. These five states together account for 50% of irrigation potential created and 45% of net area irrigated in the country. In many irrigation commands, effective irrigated area has declined due to deterioration in the distribution infrastructure. Official estimates reckon the loss of about 3 to 4 Mha command area due to water logging and salinity. Yet another study estimated that 7 Mha have gone out of farm production while 6 Mha are under increasing threat. Declining quality of irrigation service is reflected in slowing down of agricultural growth.

Ground Water

Ground water, supported by 12 million energized wells, contributing more than 50% of total irrigated area in the country has become a critical source for agriculture growth. Rapid depletion, salinization and pollution related problems threaten regions with sustainable ground water balance, whose area is continuously shrinking. Administrative blocks categorized as 'dark' or critical increased at the rate of 5.5% per annum during mid-1980s to mid-1990s. If such a trend continues then one-third of the blocks in the country would come in the 'grey' category within two decades. Groundwater mining has resulted in fluoride contamination in north Gujarat and Rajasthan and arsenic contamination in southern West Bengal endangering the sustainable livelihood of the poor. One estimate puts a quarter of India's harvest at risk from ground water depletion. In 1995, over 50% of dark blocks were located in six states namely Gujarat, Haryana, Punjab, Tamil Nadu, Karnataka arid Rajasthan.

Irrigation and Agriculture Performance in Orissa

Annual investments in irrigation sector remained consistently high as compared to many other states during the past, accounting for 20 to 25% of the state's plan outlay. Canals are the dominating source of irrigation in Orissa. Low irrigation coverage and rice dominated cropping pattern are unique features of Orissa agriculture. Poor water management resulting in low fertilizer consumption coupled with inadequate infrastructure and tenancy problems severely constrain the productivity growth in irrigated agriculture.

Scaled against ten major Indian canal commands by output impact per ha of irrigated area, Mahanadi command of Orissa ranked last. Also, in output per unit of water in the above canal commands, Orissa is at the bottom of the list with 14 kg per ha cm. This compares poorly even with the second lowest productivity (26 kg per ha cm), recorded in the Jayakwadi command of Maharashtra. This trend remained similar for different crop groups namely cereals, pulses, oilseeds and vegetables. Baseline survey conducted in selected projects in Orissa revealed that 30 to 60% of the canal command farmers cannot get adequate and timely water supplies. In fact, agricultural growth in Orissa averaged just over 1% per annum over the last two decades. Future growth in agriculture depends critically on better performance of irrigation infrastructure.

These states, accounted for 90% of the over exploited blocks in the country. Again, in these states alone, number of blocks exploiting more than 85% (dark and over exploited category) of the utilizable ground water resources has gone up from 16% to 26% during 1989-95. The magnitude and spread of over exploited blocks poses serious equity concern warranting comprehensive development and management policies encompassing all uses and sources of water. Like surface water, here too, abysmally low price regime for power neither facilitated efficiency in the use of power nor in the use of ground water for agriculture. At all India level, average power tariff for agriculture in 1998 was Rs. 0.22 per kWh, which is one-tenth of the unit cost of power supply during the same year.

Tank Irrigation System

The popular method of community-based maintenance system of tanks, existed historically, is disintegrated. Paucity of funds and meager budgetary allocations in the past resulted in continued neglect of tank irrigation infrastructure in south India, which has equity and sustainability implications. Gap ayacut (no irrigation) and stabilization areas (partial irrigation) constitute $2/3^{rd}$ of the registered ayacut in Andhra Pradesh. Within the state, only 20% of registered acute get assured irrigation in Rayalseema and Telengana regions. These regions are drought prone and account for $4/5^{th}$ of 'dark' mandals categorized by more than 85% ground water exploitation.

The linkage between tank storage and well yield in tank commands is well established. Neglected management of tank systems leads to declining storage and recharging of ground water. These factors underscore sustainability implications of deteriorating irrigation infrastructure, more so in semiarid and deficit rainfall conditions.

Tank water deficits occurs over 50% of the time period due to inadequate rainfall. Only 15% of the farms in the tank command own wells to provide supplemental source of water supply. Farms in tank commands in south India are predominantly small in size; 40% in less than 0.5 ha category, 60% in less than 1 ha size, and 80% in less than 2 ha size. Majority of the farms in the tank command being marginal holdings with tanks as the only source for irrigation and rural livelihood, deteriorating tank system has equity implications particularly in deficit rainfall situations. Currently, all 12351 minor irrigation sources taken together irrigate only 44% of the registered ayacut as against 82% in early 1950s. Loss in tank irrigated area has reached 1/4th of the net irrigated area in the state.

Similar evidences with tank irrigation systems are emerging in other major states. States of Andhra Pradesh, Tamil Nadu, Karnataka and Orissa, together accounting for 60% of the India's tank irrigated area have lost about 37% of the area irrigated by tanks during 1965-2000. There is an urgent need for rehabilitating tank irrigation system infrastructure. Physical strengthening and improvements in the inflow, storage and distribution system are needed. Water users in tank commands need to be involved in planning and implementing the rehabilitation strategies. Tanks' performance as traditional water harvesting structures, conservation and recharging of ground water besides irrigation and several other ecological functions within the villages have to be restored and their maintenance sustained.

Physical and Financial Sustainability

Providing water supplies at subsidized rates for irrigation remained the state's policy to enable secure food supplies. Currently, irrigation accounts for more than 1/3rd of states' revenue deficits. In many states, O&M expenditure was just enough for staff salaries with little for works. Low water charges and poor cost recovery resulted in secular decline in funding for maintaining water infrastructure, inefficient water allocation and sharpening conflicts over sharing of water in many regions. Current status of O&M expenditure and cost recovery in some major states, viewed in conjunction with the physical condition of the irrigation system points towards unsustainable scenario evolving in water sector, both physically and financially.

Table 1: O&M Cost Recovery, TE 2000'2

Particulars	Unit	Orissa	A.P
Potential created	Lakh ha	25	48
Gross irrigated area	Lakh ha	16	22
Average annual plan outlay	Crore Rs.	619	893
Average O&M expenditure	Crore Rs.	60	265
Weighted water rate	Rs./ha	104	398
Current water rate demand	Crore Rs.	19	116
Receipts, current account	Crore Rs.	15	69
Cost recovery, current account	Per cent	25	26

In Orissa, gross irrigated area from surface irrigation sources accounts for 64% of irrigation potential created. Average O&M expenditure remained low at 30% of the desired level. Weighted water rate, based on revised water tariff in 1998, was low at Rs. 104/ha. Current water rate demand from irrigation charges is 50% of potential demand.

All these factors culminate in poor cost recovery of 25%. Similar trends exist in other surface irrigation systems of states like Andhra Pradesh, Haryana and Gujarat. In Gujarat, actual O&M expenditure is one-fourth of the requirement. With average water charge remaining at Rs. 165/ha, cost recovery is only 33%. Similarly, Andhra Pradesh and Haryana have registered low cost recovery of 26 and 41% respectively under current account.

Physical sustainability of the irrigation infrastructure calls for need based O&M funding. This requires systematic maintenance and monitoring of the physical assets of the irrigation system and their current status on a continuous basis. Financial sustainability calls for generating the needed O&M funding from the users. And more importantly, both need to be linked. That calls for a paradigm shift in the management of water resources. Water user groups need to be empowered with the management responsibilities as well. Several states are indeed in the process of finalizing state water plans, institutionalizing farmer organizations in irrigation management and periodic review of water charges, improving assessment and collection procedures and prioritizing irrigation expenditures. Experiences so far are however mixed and the pace of progress is slow. For instance, performance of participatory irrigation management (PIM) in Gujarat indicates improved operational performance of water distribution and management. Impact on system related issues, however, is yet to be addressed. Low water rates, under assessment of irrigated area and water rate demand, and poor collection rate continue to deprive the irrigation sector from realizing potential revenue, critical for system's financial sustainability. Sustained efforts are needed for rehabilitating the irrigation infrastructure and initiating institutional reforms in water sector. Only then irrigation management transfer will become effective with system wide impacts to provide water security needed for sustainable food security.

Summing up

Vicious cycle in irrigation sector needs to be broken by empowering the stakeholders to maintain and manage the scarce water resource. Stakes are high for the users to collectively use, account and pay for it and claim their due share for system maintenance. Existing system offers no scope to integrate this process. This has implications for the sustainability of irrigation infrastructure created and added upon annually. Policy directions are needed as follows before the available options further narrow down:

- Irrigation systems (major, medium and minor) need to be restored to the satisfaction of users along with simultaneous institutional development for effective transfer of the irrigation management. Donor driven institutional initiatives obviously cannot sustain for long.

- Farmer Organizations need to be empowered to assess the irrigation coverage, revise water charges, raise water rate demand and collect receipts. Streamlining of accounting procedure to link cost recovery and O&M funding in the budgeting process is essential.

- Irrigation department should be legally empowered to identify all water user categories for broadening the revenue base and enforce quantitative measurement of water supply, charging and collection from bulk users to start with, for realizing full cost recovery.

- Any funding for irrigation development with Central assistance should be linked with mandatory institutional development as above for smooth turning over of the system to the users.

Development of institutional frameworks for an efficient use of existing and expanding supplies is central to enhance and sustain the economic and welfare contributions of scarce water resources in India. Policies to reform irrigation sector are already evolving in different states. How quickly and genuinely the institutional reforms are pursued to cover all sources and uses of water will determine India's future water and food security.

The Bank and the Big Bang

The World Bank continues to push its agenda on water privatisation even though its much-heralded examples from recent years turned out to be such dismal failures. The result will destroy countless small farmers and hand over agriculture to the rich and corporations, says P Sainath.

08 May 2005-It has been happening for some time. Maharashtra is not the first State. It won't be the last. The drive towards privatisation of water in this country was planned by the World Bank in the 1990s. The just-passed Maharashtra Water Resources Regulatory Authority Bill reeks of Bank edicts already out in 1998. In that year, the "The Irrigation Sector" report of the Bank (teamed up with the Indian Government) laid down the line.

It listed things that "need to be urgently put into practice." Among them: "drastically increasing and rationalising the current water rates." The rest of its "urgent needs" were the standard Bank rules for the capture of a country's farming by corporations. In pushing brutal hikes, the Bank was frank. Its report opposed gradual hikes. "The more recent experience is that `a big bang' approach may be better." Laughably, it cites Andhra Pradesh and Mexico as among the success stories of that approach.

The Latin American Experience

Latin America is strewn with the corpses of economies and governments that went for the 'big bang' approach. Water, especially, has been a giant factor in the rage of peoples there against regimes. This year, The New York Times ran a front-page piece on the collapse of privatised water services across Latin America. Being the Times, it coyly sidestepped any criticism of corporations. Or even of the basic concepts themselves. But it did measure the Big Bang. In Andhra Pradesh, the voters threw in a bang of their own last May. You'd think we'd learn something from all this.

Yet the new Maharashtra bill does not stray from the righteous path. It too, regurgitates the same jargon and ideas imposed by the Bank and its pet politicians and paid-for bureaucrats on the people of Andhra Pradesh and Orissa. Never mind that both States saw giant disasters in that sector. Orissa's sham (Bank-made) 'pani panchayats' shattered poor farmers in Angel district. They also handed over irrigation to a small bunch of rich landlords. In Andhra Pradesh, Chandrababu Naidu's regime passed an order that aimed for much of what the Maharashtra bill now does.

'Water Users'

In Andhra Pradesh, too, a farce of 'Water Users Associations' was set up to the applause of the Bank. Indeed, "The Irrigation Sector" report lavishly praises the Andhra Pradesh 'example.' The term 'water users' itself is intriguing. Are the rest of us non-users? Some kind of dry land bacilli? The cheers for Mr. Naidu's good example came even as his Government sold cleaned and treated water to soft drinks companies at 25 paise a litre in Hyderabad. That, at a time, when most colonies of the city were getting water for half an hour once every two days.

Meanwhile the 'users' groups proved user-friendly. They sidelined elected panchayats. The rich have always found democracy tiresome. So favoured were these groups that James Wolfensohn came all the way to Andhra Pradesh for them. To inaugurate a confederation of water users associations in 2000. He was to do this at the Koil Sagar Dam in Mahbubnagar. Alas, large mobs of angry 'non-users' furious at the loss of their water, blocked the highway.

The 'users,' far fewer in number, were given a run for their money and their limbs. Mr. Wolfensohn could not reach the site.

But if Muhammad can't go to the mountain, the mountain must go to Muhammad. The Naidu Government, famed for its efficiency in these matters, shifted the dam. In name, anyway. It took down the dam's plaque and flew it to a safe venue. Away from the ugly baying of non-users. There it had a sham of an inaugural in hiding. All this happened under the 'liberal' Wolfensohn. As against the 'hardliner' Paul Wolfowitz coming in now. It doesn't really matter, though, which Wolf is at the door, canis lupis or canis rufus. The family Canidae are predatory by nature.

How did the Bank view the mess in Andhra Pradesh? As the "remarkable strength of government commitment in Andhra Pradesh to irrigation sector reform."

Identical Jargon

Maharashtra seems set to outdo that level of commitment. This bill parrots all the pet phrases of the Bank. It dittos the ideas, rules and structures that the Bank's own vision lays out. In parts, the jargon is near identical. But it breaks some new ground. 'Entitlement' in this bill is not defined as the right or claim of a citizen or community. Here it means 'any authorisation by any river basin agency to use the water for the purposes of this act.' In short, the entitlements of authority, not of society, are what drive the bill. The bill also equates private companies with citizens. The section on State Water Planning is clear on this. "The expression 'person' shall include individual, group of individuals, all local authorities, association, societies, companies etc.," In short, petty officials and giant corporates will have the same rights as citizens and farmers.

Huge Costs Involved

It warns that in some regions, "Water shall not be made available from the canal... " Not "unless the cultivator adopts drip irrigation or sprinkler irrigation..." Or whatever the authority orders. This could add Rs. 15-20,000 per acre to the farmers' costs for just installation. Running costs would be a further burden. This is a rip-off. Well-known private companies close to the ruling outfit will strike gold. The State might even buy this equipment from them in the name of subsidies to the farmer. Even if the farmer cannot cope with running costs.

The new Maharashtra Water Resources Regulatory Authority "shall consist of a chairperson and two other members." The chair will be of Chief Secretary rank. Of the other two, one "shall be an expert from the field of water resources engineering." The other, likewise, "in the field of water resources economy." There's another open door for the private sector-right on the top floor.

This body will ensure that "water charges shall reflect the full recovery of the cost of irrigation management, administration, operation and maintenance of water resources project." Also hidden in the deal is a clause that sailed through when the bill was first passed by the Legislative Council. That talks of partial "recovery of capital investment."

These levels of cost recovery are aimed at clearing the way for private investors. The Maharashtra bill, as economist and former State Planning Board member H.M. Desarda points out, could make costs unbearable. Perhaps as much as Rs. 8,000 an acre. That would simply evict lakhs of small holders from farming.

Some of those who back the bill, like MLC B.T. Deshmukh, point out that it gives priority to backward regions. That new projects must come first to hard-hit Vidharbha and Marathwada. True, the terms of the bill do imply this. And so? It's like if the Bombay Gymkhana were to give first preference in membership to those living in the slums of Dharavi. Sure, they'd get priority. Could they afford an 'nth' of the charges?

Two-child Norm

The uproar on the bill centred around the obnoxious two-child norm. But that is just the tip of the iceberg. On April 28, M.P. Veerendrakumar drew the Lok Sabha's attention to The Hindu's reports on the subject during a discussion on the Finance Bill. "Marginalised farmers and those who take agriculture as a livelihood will be driven out. The field will be entirely open for big tycoons and MNCs... " "Whenever agriculture issues are raised in the House," he argued "the reply is that it is a State subject." But he points out, "the moment some bureaucrat goes to some country, he signs an international agreement." With whose authority, he demanded to know. He believes a constitutional amendment is needed to root out the secrecy, intrigue and plain old corruption that are tied with such legislation.

The distribution of water already stands privatised in parts of several towns across the country. But applied to farming, will it work? Can such massive rates be recovered? Absolutely not. No one can pay. So why bother, then?

Because it will destroy countless small farmers. It will establish, yet again, water as a private good not as a human right. (What impact the costs will have on food prices has not even been looked at.)

It will hand over agriculture to the rich and corporations. It will worsen the terrible situation of poor farmers in the State amongst whom there have been hundreds of suicides. And it will doubtless be touted as a national and global 'model.' Watch out for that big bang.

Socioeconomic Issues in Irrigation Literature: Approaches, Concepts, and Meanings

When engineers talk social sciences, what do they mean? The aim of this work is to analyse the presence of social science topics in the recent irrigation literature. But, what is the point-the reader could ask-of looking for socioeconomic subjects within a scientific production mostly destined to engineers?

The "Knowledge Assessment on Sustainable Water Resources Management for Irrigation" (KASWARMI) project has found some scientific and technological niches that ask for new research on the "technical" aspects of irrigation in pursue of sustainability, efficiency, productivity, lower costs, ecological sustainability, etc. Also, the project has identified a number of consolidated technological innovations or improvements for making irrigation closer to sustainability that have not been adopted by agricultural water users along Latin America. But further than this, the project results have made evident many cases of unsuccessful or non sustainable irrigation experiences that could only be explained by the underestimation of deep socioeconomic issues. This involves not only the social factors related to the implementation/adoption of new or better technologies by a wide spectrum of users but also the way in which water use and irrigation projects are conceived, planned and implemented by scientists, politicians and practitioners. In other words, sometimes all the improvements already achieved in the "technical" aspects of irrigation seem not to be enough when socioeconomic factors have not been carefully addressed in each and every stage of irrigation projects. At this point, social subjects become a matter of attention for irrigation engineers and the KASWARMI project was interested in assessing the social science inputs irrigation specialists receive.

From this point, the starting hypothesis of the review was that social science issues are not present as it would be desirable and thus not sufficiently valued, and that the approaches to social issues in this field lack comprehensiveness. Beyond any assumption made, the review also aimed at exploring the particular ways in which social issues were approached in this literature and linked with the more traditional subjects. With these objectives, the paper presents a state of the art on selected social subjects as they appear in the irrigation literature, and analyses the way in which they are conceived, thought and articulated with the more "technical" factors of irrigation. This should be useful to raise new avenues of research and to enhance articulation of "technical" and social science approaches in quest of a more close to sustainability irrigation practices. In the context of the project agenda, this state of the art-along with detected needs of Latin American stakeholders-was an input to identifying gaps of knowledge towards sustainable irrigation.

Covering the most relevant of social science issues that aroused within the project discussions, five main subjects were investigated: a) Conflicts around water; b) Equity among users; c) Actors and stakeholders around water; d) Cultural background; and e) Gender.

The time scope for the search was established in the last seven years, from January 2002 to December 2008, though some subjects demanded going back beyond 2002 for finding a minimum of papers to analyse. A first search included eight of the most prestigious journals devoted to irrigation subjects: 1) Agriculture Water Management, 2) Irrigation and Drainage, 3) Irrigation and Drainage Systems, 4) Irrigation Science, 5) Journal of Irrigation and Drainage Engineering, 6) Journal of Water Resources Planning and Management, 7) Water Resource Management, and 8) Water Resources Research. After the search within these journals yielded meager results in terms of the amount of papers found, a new search had to be conducted, this time without restraining to specific journals but moving freely with a deliberate thematic purpose.

Review

In spite of the amount of papers reviewed (more than 5000 just in the eight journals mentioned), those related to the studied subjects were extremely scarce. That was the first finding. Papers appeared to be more numerous when the search was extended to diverse social science subjects and when it reached other sources, but then they appeared mingled with a diversity of social science topics and scattered through an ample variety of social science editions, in which journals were only a portion. This situation implies difficulties for irrigation specialists to even be aware of their existence, not to mention to be familiar with them.

Even if it was complicated to separate the bibliography about water conflicts, the actors and stakeholders involved, the equity issues and the cultural and gender topics, the review was organized in five thematic cores:

Actors and Stakeholders around Water

Some of the papers that analyse the actors involved in water management approach the issue from the study of the water-governance point of view: Who are involved in water policy processes? How do they interact? How is the State playing its role? How are users organized? Scholz and Stiftel (2005) answer these questions by trying to provide clues to management. In his analysis of the governance of irrigation systems, Palerm-Viqueira (2007) differentiates government from management and explores instances of self-governance. Rogers (2002) presented the fundamentals of water governance in Latin America and exemplifies with cases from Mexico, Brazil, Chile, Argentina and Honduras.

Concerning the actor typologies from which the problems relative to management of irrigation systems have been addressed, the analysed literature poses a traditional division whereby those actors representing the State are distinguished from others that belong to the "self-management" category. Although this division has been commonly applied in the analytical field, and a proof of this is the fact that it has been supported by numerous authors, in the particular case of Latin America, there is a need for nourishing this scope by introducing a new category. This is to distinguish within the self-management group those social actors with a unified authority, who act as a specialized group within this frame, from those other systems where the actors operate on the basis of their own knowledge, both to distribute the resource and to organize the tasks inherent to the system and solve the problems generated in relation to these actions.

The growing presence of economic capitals over the traditional farming scheme in Latin America is present in the actors/stakeholder analysis. If, on its part, the State functions as the epicentre of the thematic issue and centre of legitimate authority, other actors render the panorama more complex by bringing onto the stage the equity issue. Irrigation systems, and more specifically the management of the water resource, acquire political edges, and water becomes an arena of dispute where there are actors from the State, from different civil associations organized on the basis of groups of experts and, moreover, from original peoples who claim for the application of sustainability criteria that engage in open competition with those supported by the former stakeholders.

Also related to the actor typologies, the irrigation water user is the most frequent mentioned and analysed stakeholder. Although present in some analysis, other water users are less frequently considered. It is interesting to refer to Molden *et al.* (2007) concerning included and excluded actors when establishing management specifications for pro-poor irrigation services. They pointed out the convenience of considering multiple uses and users of water and to give due attention to the many other people dependent on irrigation water including the landless, livestock keepers, fishermen and domestic water users.

Along with governance, the participation issue gathers a good deal of the discussion about actors and stakeholders around water. Participation and community involvement in water users management are the object of some papers, most of them presenting experiences in cases such as a canal management in Uzbekistan (Abdullaev *et al.*, 2009) and building and managing temporary check-dams in southern India (Balooni *et al.*, 2008). Singh *et al.* (2008) addresses the participation issue presenting the usefulness of water users involvement to enhance irrigation projects sustainability once the financial or technical initial support has been withdrawn. Participatory processes are also considered a lever to sustainability by Maleza and Nishimura (2007) when analysing the national irrigation system management in Bohol, Philippines. Going back to Singh-and in tune with our assumptions-, the paper signals that perspectives of local people's needs are crucial to the development of research and extension efforts, which would also "help researchers and practitioners to make better choices and more informed decisions when designing their research, communication and dissemination approaches".

Users' perspectives are also taken into consideration for improving the performance of water users' organizations involved in poverty alleviation initiatives in the Fergana Valley, Central Asia. The participation subject in the context of the stakeholder-researcher cooperation is analysed by Ritzema *et al.* (2007), in this case as part of the research method. Ounvichit *et al.* (2008) presents the social relationships between individual farmers and their communities as "a promising scaffold for water users' organization" while Vandersypen *et al.* (2007) proposes didactic tools for participatory water management that could support water users' associations in coping with their responsibilities after the withdrawal of the State.

Great part of the revised papers deals with the identification (definition) and analysis of the actors and stakeholders involved in water management processes, either more or less conflictive. Some of these papers, especially those with a sociological background, understand the management of water-and of irrigation in particular-in terms of actors, stakeholders, practices and power, revealing issues related to social interplay and to the mechanisms and conditions producing and reproducing material and symbolic relationships of social dominance.

Conflicts around Water

Although a variety of lines of thinking were found around water conflicts, almost all coincide in two shared elements. On one hand, water is considered a scarce good and, in this context, conflicts arise as a water shortage crisis and the subsequent water use concurrence. On the other, water is a common good exposed to private appropriation. Further than that, approaches to water conflicts diverge.

A first line of thinking found in the literature is built around the ecological sustainability of water uses. Under this point of view, conflicts arise when an intensive water use compromises the ecosystem integrity. These concerns are thoroughly developed for the global scale. The 2ⁿᵈ United Nations World Water Development Report: "Water, a shared responsibility" is an example of this. Its Section 2: "Changing Natural Systems" presents an overview of the state of water resources and ecosystems and explores current assessment techniques and approaches to integrated water resource management (U.N., 2006). Bos (2002) of the International Union for Conservation of Nature (IUCN) argues that ensuring water security requires integrated management of water resources, by balancing between natural and human needs at the ecosystem level, and by accounting for the actual value of natural services in development decisionmaking.

Also in this global scale, another group of papers refers to water resources as a mayor geopolitical problem, even as a potential war space where already existing problems could break out and new ones are expected for this century. Some authors identify "neighbour conflicting countries" or "transboundary conflicts" and even "terrorism of water resources. There is even a special issue of Journal of Water Resources Planning and Management (2007) on "Transboundary Water Sharing" with different approaches: historical perspectives (Phelps, 2007); conflict (Matthews and St. Germain, 2007); governance (Draper, 2007); water management and water laws (Dellapenna, 2007); effects of climate change (Draper and Kundell, 2007), among others. Some other authors discuss water conflicts between rich and poor countries (Hoekstra and Chapagain, 2007). The first group assesses water war risk identifying even "hydrostrategic territories" (Wolf, 1996). The second identify countries-stakeholders in a conflict caused by unequal distribution.

In some papers, water conflicts express local-global relationships. This is the approach developed by some papers analysing the local effects (upon water resources) of economic globalisation and global climate change upon water resources availability, allocation, management, etc. Within the first group, there are some papers on the commoditisation and subsequent privatization of water, denouncing the advance of economic powers over a strategic resource and the introduction of water allocation market mechanisms. Here, the water wars are between citizens and corporations (Shiva, 2001).

In this context, some authors and also global institutions present water as a basic human right, nurturing a human rights-based approach to water. Some of these "conflict studies" lines have the correspondent "conflict resolution" papers, particularly those related to local conflicts: some propose prevention (Wolf, 1996), incentive-compatible cooperation strategies, consensus based mechanisms or a better water management for conflict resolution, among others. There are also "conflict resolution tools" of different nature: law based multi-criteria decision tools; decision support systems; game theoretical concepts; graph models and database tools. Not surprisingly, improvement of water managements systems is presented as the clue for solving an ample spectre of irrigation conflicts and beyond, the alleviation of rural problems, poverty in the heart of them.

With respect to those that look at water conflicts in the context of global climate change, Rojas *et al.* (2006) argued that power differentials in water conflict resolution between stakeholders may increase the exposure, hamper the adaptive capacity and therefore increase the vulnerability of communities to global change. These authors point out the utility of analysing water conflicts as they can provide insights that can be applied to understanding adaptive resolution of water conflicts and offer important institutional and social learning for adapting to future climate change-induced water conflicts.

On more delimited scales, a good deal of the conflicts referred in the literature are related to irrigation practices, as clashes between users of irrigation water, differences between agricultural users and other users (water for human consumption, industrial, recreational uses, mining uses, etc.). Rajabu and Mahoo (2008), for instance, presents a conflict solving tool based on participation of stakeholders and analyse its application in a sub-catchment in Tanzania.

Finally, a special chapter on water conflicts is developed around what could be called social effects of dams, these considered as mayor pieces of the irrigation systems. Supported by technicians, government officials usually say these dams and an extensive irrigation system will bring electricity and water to areas suffering from drought. Arguing that the benefits are exaggerated and the costs underestimated, a great deal of papers describes and analyses the effects of big dams on vulnerable social groups, especially on aboriginal peoples, and those causing relocations. On these papers, the water conflicts are expressed in terms of equity and cultural struggles, themes that will be analysed in detail later on. "The colonisation of rivers" is the way in which Shiva (2002) refers to dams as associating them to water wars.

To sum up, water conflicts arise as a dimension built over struggles for a scarce resource as an arena where competing interests clash: water access problems, allocation disputes, availability, security and sustainability issues, etc. Interest conflicts and opposing points of view disclose different rationalities and particular cultural backgrounds at stake, attesting that cultural background, equity and gender turn out to be cross-subjects.

Equity Among Users

At first, the equity issue appears in the irrigation bibliography as distributive conflicts among economic sectors (e.g., agriculture vs. industry), among users within the same sector (e.g., farmers vs. peasants), among countries, regions or places sharing a common source as in the typical case of the upstream-downstream conflicts (Gaur *et al.*, 2008), among urban and rural users, indigenous groups and modern communities, rich and poor, men and women, and even among present and future generations. Phansalkar (2006) distinguishes and defines social equity, spatial equity, gender equity and inter generational equity. Wilder and Lankao (2006) and Moyo (2005), on their part, analyse the inter generational equity specifically. But in many papers the concept of equity is often undefined and usually ambiguous as Wegerich (2007) argues as a prologue to the exploration of aspects of equity of water allocation between different riparian states and districts in Uzbekistan. Further than these uncertainties, the general consensus is that there are differential in access and appropriation conditions for different users. For the case of upstream-downstream conflicts, for example, Van der Zaag (2007) recognizes asymmetries for Southern Africa and addresses to the institutional arrangements that can be devised to (re-)establish an equilibrium between up-and downstream entities within a catchment area or river basin. It is also van der Zaag who proposes the concept of "hydrosolidarity" as "a normative value that may help to recreate the balance between the various (asymmetrical) interests that exist within a river basin".

In another significant number of papers-referred to different territories of the world-the water-equity concern is expressed in concerns about the impacts of markets mechanisms and property systems. Miller (2004) wonders about the objectives and effects of water reforms, more inspired by physical and technical objectives than by governability and equity issues. He also argues that in some cases, the losses and inefficient uses of water yield benefits to ecosystems. In a more specific approach, some authors wonder about the way in which the prizing of water impacts over a variety of stakeholders, each one affected by different situations. On the same line of thinking, Manos *et al.* (2006) simulate the impact that various policies based upon the water price have on agricultural production and analyses the economic, social and the environmental implications of alternative irrigation water policies using a multicriteria model.

In reference to equity and water conflicts induced by the implementation or modernisation of irrigation systems, a series of papers explore the question of whether the improvement of the traditional irrigation systems bring benefits in terms of equity and in reduction of water conflicts. Some authors discuss the hypothesis of traditional systems improvements bringing more water to rural poor and thus mitigating inequities. For the case of a smallholder irrigation system in Tanzania, Lankford (2004)-for example-indicated that the improvement of the system does not necessarily result in improved water performance, greater equity and reduced conflict. The usual outcomes of such projects-he argues-is a gain in water for the system being upgraded, especially if located upstream, accompanied by less ability to share water at the river basin scale.

In another paper, an irrigation improvement programme of modernization with a structured system concept is analysed by Sakthivadivel *et al.* (1999) for the Bhadra Project in India. They found that although agricultural productivity has not registered a significant decline since before the intervention, preferential allocation to head end of command continues and inequity sets in within the distributary commands. The tail-end water supply deprivation is partially offset by farmers practising deficit irrigation.

Farmers' organization and participation in decisionmaking at scheme level and water distribution at distributary level and below are very low. With the same concerns but in a working line focused on the development of analytical tools, Thiruvengadachari and Sakthivadivel (1997) have usedinstruments such as satellite remote sensing, geographic information system (GIS) techniques, and hydrologic modelling to assess the same Bhadra Project in India. Spatial and temporal information has helped analysts evaluate the performance of the agricultural system over several years and across the irrigation scheme. The results have shown significant improvements in agricultural productivity while confirming equity problems (The equity of the water supply is measured here through Christiansen's Uniformity Coefficient). For the case of the China's lower Yellow River basin, Roost (2003) introduced a new irrigation model (OASIS) that allows proper quantification of water use efficiency, productivity and equity under actual or hypothetical conditions of land use, infrastructure and water management.

Finally, it should be noticed that water equity issues have not only poverty implications but also gender and indigenous dimensions that will be mentioned below.

Cultural Background

The issues of water management and distribution and further on, the concerns of irrigation systems for their sustainability, has captured the attention of the scientific sector. However, a meticulous analysis of the existent literature shows that the orientation followed as well as their guiding hypothesis and supporting theories differentiate each other according to the disciplinary background of their authors. From the agronomy, and engineering sciences in particular, irrigation systems have been analysed with higher emphasis on the quality, quantity and productivity of the water resource, or as means to operate improvements in these senses. In the case of social sciences and particularly of anthropology, the concerns have been oriented to show instead.

That water constitutes an asset that exceed its immediate materiality and that integrate symbolic dimensions That at the same time it shows a concrete materiality, water and irrigation schemes constitute channels that facilitate, promote and even explain forms of organization characteristic of certain social groups. That means, water is the base of social relationships, generate forms of organization and at the same time show the non-material dimensions that are also part of the actors' social life.

In the case of agronomic and engineering sciences, as pointed out before, it can be observed an affinity to think about the more material aspects of the subject, furthermore it is necessary to notice that since some time ago, there are increasing concerns about equity in distribution and participation in management, because of conflicts about management that have emerged between actors. The knowledge emerged from the scientific sector about irrigation management came from sources linked to social sciences and the data, to which is possible to have access correspond to a varied casuistry dispersed worldwide in which it can be observed a bigger Asiatic presence.

Reviewed bibliography show that the valuation of the cultural background is mostly related to traditional knowledge and linked to the casuistry considered. It is assumed that water is a public good and a valuable resource that is not to be wasted. Based on the former, arguments are formulated in favour of the sustainability that traditional knowledge bears in relation to the management of water for irrigation as well as for human consumption. On the other hand, it is stated that modern agricultural development efforts often ignore this indigenous knowledge, replacing traditional infrastructure with new construction, and replacing indigenous management arrangements with state bureaucracies (Groenfeldt, 2005) undervaluing what appears to have been quite productive and sustainable before extra-cultural influences began (Cleveland *et al.*, 1995).

Because of it, it is also sustained that indigenous irrigation systems should be intelligently assisted, rather than mindlessly replaced. Coherently with the former, Varisco (1991) stated that farmer knowledge can contribute to sustainable production and can be grafted on to modern methods and technology. In this context it is stated the growing interest in using the traditional knowledge, which should be captured to aid in propagation of cultural methods of production and associated technologies (Gillespie *et al.*, 2004).

Gender

Papers on this matter were not numerous. Two papers, one on an overview on gender and irrigation and another approaching gender within the diversity issues (Hussain, 2007a) are part of the special issue of Irrigation and Drainage journal: "Irrigation and poverty alleviation: Pro-poor intervention strategies in irrigated agriculture".

Gender approach has increased its importance in the last decade. It includes the analysis of relationships between agricultural systems and the responsibilities and rights of male and female farmers, according to the local agroecological and cultural context in which they develop their (agricultural) activities. Nevertheless, these issues are not dealt in "main stream" publications about irrigation and water management but within policy and development studies, over all those promoting a reflexive approach to state interventions.

However, the increasing demand about participatory planning in agricultural sector is still far from being covered by practical solutions, to accomplish objectives of minimizing differences in socio-economic, cultural and gender terms (Koopman, 1997). Those objectives consist not simply in (explicitly) including women in public and development policies (for instance in agricultural or irrigation policies). The approach analyses the different roles and responsibilities of both women and men, by recognizing their differences in access and control over resources, and therefore the consequences, conditioning factors and difficulties to reach such a goal.

The literature review shows a critical situation of water allocation and rights as a result of water scarcity and at the same time because of intervention programmes aimed to increase the efficiency of water allocation and delivery. The studies also underline how policy and irrigation planning have mainly focused on construction and maintenance of irrigation infrastructure, irrigation efficiency, water productivity as well as the evaluation of the effects of agricultural and irrigation practices on soil, ignoring the needs and priorities differentiated by gender as well as the nature of the cultivated products, the impacts on labour markets or the coexistence of multiple uses of water (for production or consumption).

In the analysis is underlined how irrigation might, eventually, contribute to food insecurity, because of the trend to modify agricultural patterns that involved local knowledge and farming practices, soil management practices, etc., that are also replaced by new (cash) crops and technologies for export. There are evidences how children coming from cash-crops farms are poorly fed in comparison with those coming from the so called "traditional farms" producing a diversity of staple crops.

There are no doubts that access to irrigation water constitutes not only an important asset, but also a source of power and conflict. In this sense, organizations promoting a gender approach vindicate the need to strengthen participatory spaces for capacity building and communication, oriented to create incentives for the different expressions of rights, duties and social inclusion. Some examples are those showing the importance of participation of different stakeholders (including women) in decision making, contributing thus, towards a more sustainable irrigated agriculture and water resource management and conceiving irrigation as a social construction.

Along with Singh *et al*. (2004), claiming for an holistic perspective on water management (in this case for domestic water supply systems), its worth to remark, as the other side of the coin, the need of objective improvements in the situation of women; because it is not always clear that the actual effects of those changes will improve in their social, familiar and personal situation. This is why Singh argues "the need to design participatory paradigms that are more realistically rooted in community-based institutional frameworks so as to enhance effectiveness of the endeavours" (Singh, 2007).

Evidences show the difficulties in positioning the gender issue in the main discussion related to sustainable irrigation. When the discussion is included, it is done from a very specific, sector-oriented or institutional perspective. However some tools have been developed (for instance by FAO), to be used by irrigation engineers, government organizations and non-governmental organizations (NGO) to improve intervention projects by including perspectives from rural women or other disadvantaged groups. Finally, it is important to state that in this respect, two perspectives converge. On the one hand, those promoted by donor and international agencies and special services from world agencies, which have developed the main documents to approach development from a gender perspective (FAO, International Union for Conservation of Nature, World Bank and others); on the other hand, intervention agencies such as NGOs, which promote the linkage with local groups. Between these two perspectives, it is still incipient and very limited the capitalization of those approaches related to the production of scientific knowledge on irrigation and the application of this knowledge among irrigators.

Others Issues

Within a variety of social issues related to water and irrigation that were found in addition to those mentioned above, poverty showed to be a recurrent subject linked to irrigation impacts. A major contribution to issues linking irrigation and poverty (or fight against poverty) is made by a special issue of Irrigation and Drainage journal: "Irrigation and poverty alleviation: Pro-poor intervention strategies in irrigated agriculture". Some of the papers on this issue present results and insights coming from the International Water Management Institute (IWMI) studies and projects on Asian cases. They explore the relationships between irrigation and poverty, and particularly the irrigation initiatives as poverty alleviation strategies. Some papers summarize the results, conclusions and lessons learnt from cases of pro-poor interventions.

Others approach poverty alleviation strategies through reforms in irrigation water rights (Bruns, 2007) or irrigation management reforms (Wang *et al.*, 2007). Molden *et al.* (2007) calls attention to performance assessment in irrigation for poverty reduction, while Namara *et al.* (2007) introduced land issues and gives insights on land and water management innovations. It is also interesting the analysis of Narayanamoorthy (2007) on the nexus between groundwater irrigation and rural poverty, stating that access to groundwater is a poverty protection factor. But groundwater use is not a panacea as Llamas and Martinez-Santos (2005) identify it as a potential source of social conflicts.

Finally, it is relevant to bring up that a few papers show concerns about the role of science in contributing to social benefits through improvements of irrigation projects and practices. Inasmuch as experiences of implementation of irrigation systems show that a good part of the problems originates within the social field, the development of irrigation in the real world imposes some demands on the scientific sector. This dilemma about the direction of the scientific development is addressed by Shuttleworth (2007) when he wonders about a "Stakeholder-driven, enquiry-driven, or stakeholder-relevant, enquiry-driven science?" Apparently simpler but not least important is the concern about available river basin management insights and information not being of help for water managers.

Pahl-Wostl and Borowski (2006) realized that simply providing information does not result in an effective communication across the science-policy interface. In addition to Pahl-Wostl, other authors contribute to this Water Resource Management special issue: "Methods for Participatory Water Resource Management". Borowski and Hare (2006) identify a gap between water managers and research community that is evidence of a mutual misunderstanding of the fundamental activities of both communities, while Brugnach *et al.* (2007) refer to computer models pointing out troubles for integrating the information derived from models into policy.

They partially explain this situation in the lack of confidence policy makers have on the incorporation of modelling information into policy formulation; they examine the reasons for this apparent lack of confidence and explore how some tools, presently in use, address this problem. Beyond this special issue but related to the subject, Keuls (2008) approaches the issue from a capacity building point of view asking for a knowledge network development for the water resource management sector, while Maguire (2003), for a case in USA, identifies the most serious shortcomings resting not with the scientists or the stakeholders, but with the too narrow structure of a regulatory process unable to encompass the stakeholders' wide-ranging concerns.

Results

In spite of the amount of papers reviewed, those related to the studied subjects were extremely scarce. More papers were found scattered on social sciences editions, but far from the reach of irrigation specialist and not always close to implementation purposes. In the context of this meagre presence, it looked like social issues tend to appear more frequently in the recent years suggesting an increasing interest in the socioeconomic dimensions of irrigation, although a sample limited to a seven years search is not enough to establish a clear trend.

In addition to the five specific social science topics deliberately looked for, poverty appeared as a recurrent concern. The role of science and the relationships between scientist, policy makers, decision makers and other stakeholders was also present in some papers.

Cases in Latin American countries seem to have a minor representation in relation to developing countries in other continents, particularly Asia. The same happens in terms of language, being most of the scientific production written in English language and in much less proportion in Spanish. This may constitute a barrier for the access to scientific knowledge by Latin American researchers, apart from technological and financial barriers to access such literature.

Social aspects of sustainable irrigation are not considered or just as a "context" issue in classic irrigation journals. When the five specific social science topics of this state of the art appeared in the universe of the selected irrigation journals, it can be observed that in most of the papers they are not the key issues but rather side topics related to problems that focus on the "harder" aspects of sustainable irrigation. For instance, in papers making their contribution to multiple attribute systems, integrated management or planning or multi-agents, what is being afforded to the traditional engineering view is either environmental issues that attempt to ensure or facilitate the ecological integrity of the water system, or economic factors that internalize costs not previously considered, bring transparency to subsidies or assess the situation of different stakeholders with higher or lower payment capacity in systems pointing to "economic sustainability".

The journals selected for this search are highly prestigious and widely consulted by those involved in studying and practicing a more sustainable irrigation. They show a particular bias as they prioritize the most technical sides of the problem: channel, pipes, reservoir, barrage design and calculations, different irrigation technologies, irrigation performance, efficiency, infiltration, evapotranspiration, aquifer performance, groundwater flows, etc. A marked orientation toward action-typical of engineering disciplines-is, on the other hand, observed, as well as the will to spread the advances relative to tools, techniques and methodologies: equations, algorithms, coefficients, matrix, formulas, numerical simulations, models, etc. Social issues are not yet the concern of these publications and are rarely addressed. And when they are, the overall purpose is usually to contextualize technical hydraulic or agricultural problems or to give a reference frame to their implementation. It could be concluded that the treatment of socioeconomic factors lack comprehensiveness, as they don't appeared to be fully articulated with the technical subjects of irrigation, at least in this type of publications.

There is a trend to discuss social issues as emerging elements from the actual practices and intervention processes, being that such process were initially conceived from a technical and engineering point of view. Therefore, most of the literature dealing with the five subjects discussed is reported as experiences though case studies. In many cases water and irrigation is directly associated to the notion of development as a desirable and ideal situation, and embedded to the notion of "progress", as a component of "civilization" and, in an opposite direction to "savagery" and as a mean of transformation towards a capitalist and modern economy, including and "integrating" cultural minorities and ethnic groups. It is remarkable that the studies under this perspective, instead of promoting exhaustive analysis of such complexity, tend to define the studied situations as a traditional, or "defective" situations (underdeveloped, backwardness) and also omitting in the analysis, the multiple dimensions (economic, cultural and political) explaining social practices (for instance agricultural or irrigation practices).

Related to the previous discussion, it is important to state that, beyond the material and technical dimensions associated to water and irrigation, extra-economic and symbolic dimensions have been approached mainly from anthropological perspectives and in a less extent from other disciplines.

Even though there is strong evidence that socioeconomic issues are at the base of a good deal of unsuccessful or non sustainable irrigation processes, socioeconomic subjects are not frequent in the irrigation engineering literature. When they appear, they are often reduced to references to contexts more than an object of study by themselves or an input for decision making.

Although disciplinary biases and preferences are perfectly reasonable, omissions in this field may result in serious risks to sustainability of intervention processes. A desirable trend would be that social dimensions of irrigation would be incorporated in scientific knowledge according to the current demand in practical situations, strengthening thus the development of new approaches to irrigation management. But irrigation engineers cannot be held responsible for not reading social sciences issues as scientific editorial lines-and moreover-scientific research trends often lead to ever more specific knowledge.

The situation, as seen from this state of the art and from the KASWARMI project results, lies in an irrigation science and practice getting increasingly complex and progressively more concerned about sustainability. The notion of sustainability requires thinking economics, social and environmental as three interrelated dimensions, posing a challenge to science and scientific literature, as they have to achieve the frequently proclaimed interdisciplinary approach. It also defies practitioners, as irrigation projects need engineers interested in social issues working together with social scientists willing to involve themselves in engineering projects.

Paddy Irrigation and Water Management in Southeast Asia

The evaluation covers six gravity irrigation schemes with reservoirs for water storage in Myanmar, Thailand, and Vietnam. Widely dispersed in the region, the schemes were chosen for their variety, rather than their typicality. Though paddy predominates in all, four of the schemes are large—at least 40,000 ha—and the other two are small. Four have plenty of water; the other two, in the central dry zone of Myanmar, have much less than planned. To compare the organization and effectiveness of operation and maintenance (O&M) between irrigation and flood control, the study also reviewed the performance audit findings from flood control schemes at three sites in Bangladesh.

The evaluators visited farmers and officials at the scheme sites, and public irrigation authorities responsible for the schemes. Interactive group and household interviews were arranged in all four countries. The evaluation focused on agroeconomic impacts, and on operation and maintenance issues.

Agroeconomic Impacts

In the Irrigation Schemes

Crop yields have been close to expectations:

- In four schemes, including three of the large ones, irrigation covers much smaller areas than planned. The main reasons are planners' over optimism, engineering errors, lower-than-normal rainfall, and failure to extend the tertiary canals.
- Cropping intensities are much lower than expected at three sites and are falling at a fourth.
- Only one scheme—the small Azin scheme in southern Myanmar—has reached its targets for both area and cropping intensity.

- In two thirds of the schemes, output is much less than expected when the projects were appraised. Output of paddy, and of a few other major crops at the two schemes where paddy is not completely dominant, is from 32 to 73 percent of appraisal estimates in five schemes. Again the exception is Azin in Myanmar.

- The collapse in international rice prices since the early 1980s, after the projects were appraised, helped drive down profitability. In real terms, the price of rice in 1995 was only one-third of the 1995 price that was projected in 1980. Farmers could have coped more easily with the price decline had they diversified their cropping pattern, as planned in four of the schemes. Instead, they grew more paddy.

- Economic rates of return are no higher than 7 percent in all schemes and are negative in one.

By the Bank's standards, guided by opportunity costs, these have been uneconomic investments. But borrowers point to the visible signs of substantial intensification of agriculture and increases in yields over large areas that were previously rainfed—as well as the considerable indirect regional and social benefits of the investments that are not captured by rate-of-return estimates. They are committed to maintaining the schemes.

Welfare Impacts

In most cases, incomes from paddy in the project areas are well below appraisal estimates. In Vietnam and Thailand, actual incomes are only 10–30 percent of appraisal estimates. The gap is lower in Myanmar, but mainly because appraisal projections were less ambitious. Farm incomes in the project areas may not be high enough to keep families, and especially their youth, committed to farming. Net household incomes from irrigated cropping on average-size farms in the six schemes range from $600 to $2,000 a year, depending on the size of the farm, local market prices for paddy, and the extent of diversification out of paddy.

Unless these farms adopt higher-value cropping systems, their sustainability is in doubt. The relation between low paddy prices and rising wages suggests an irreversible trend. In Thailand, the world's largest rice exporter, economic modernization is pulling farmers into factories even as low paddy incomes push them out. Vietnam and Myanmar, also rice exporters, can be expected to follow the same route.

Operation and Maintenance

Through field surveys, the evaluators assessed the performance of agencies and farmers in operating and maintaining the schemes. They observed the condition of canals and control structures, agency activity in allocating, distributing, and maintaining the flow of water, and the strengths and weaknesses of O&M by farmers. In the irrigation schemes, they found that:

- Water appears to be generally well managed. In the three major schemes, water-use efficiency ratings are high, at 43–52 percent. This compares with the 25–40 percent ratings that the International Food Policy Research Institute quotes as typical in Asian irrigation schemes.

- Though errors were made in design engineering, there is no evidence at the six sites that original construction was so poor as to make maintenance difficult.

- Agencies generally operate and maintain irrigation structures quite well, at least as regards operational plans. Dams, main canals, and structures on these canals under agency responsibility are in good order, except that many measurement devices have been removed or allowed to deteriorate. In Vietnam, the knowledge and hands-on involvement of field engineers from the two provincial irrigation services at Dau Tieng; in Thailand, the purposive reforms promoted by the Royal Irrigation Department's new project manager at Lam Pao, backed up by technical assistance; and in Myanmar, the intelligent management of scarce supplies by the Irrigation Department at Kinda, are impressive.

- Sophisticated measurement and allocation programs promoted by consultants have been abandoned, both in Myanmar and Thailand. They are premature for paddy cultivation, given the loose institutional arrangements in place and the high water tolerance of rice.

- Assessed against engineering design, irrigators behave poorly. Assessed against their collective self-interest on individual watercourses, their behaviour makes sense. In all the schemes surveyed, farmers maintain their canals to serviceable standards. Vandalism and neglect affect mostly structures that are ill suited to community needs, such as tertiary gates that interfere with the flexible operating protocols favoured by farmers, inlets that induce excess flooding in the lower reaches, and embankments that prevent drainage. When siltation and weed infestation threaten irrigation, farmers readily mobilize labour to clean up watercourses. They handle routine maintenance and minor repair of tertiary canals, channels, and associated structures on a collective basis with a modicum of support. The evidence confirms that farmers will act collectively in the common interest if substantial benefits, broadly available to the irrigator community, are at stake.

· Farmers share water. Crop yields differ little from the heads to the tails of watercourses. Farmers' own customary arrangements allow significant advantages to those at the heads of canals, but provide enough water for those at the tails. Even where water is short, relations between farmers at the heads and tails of canals are quite civil and accommodating, suggesting that enough "social capital" is available to overcome collective action dilemmas.

Role of Water user Groups

Water user groups (WUGs) are not functioning as well as expected. Nominally, water user groups exist throughout the command areas of the six irrigation schemes, and are responsible for O&M below the turnouts of tertiary canals. Public agencies retain ultimate responsibility for operating and maintaining the tertiary gates, but usually share these jobs—especially the operation of gates—formally or informally with unfederated tertiary water user groups.

The water user groups vary in type and effectiveness. In the internationally assisted sections of the scheme at Lam Pao, Thailand, both the water user groups and the federated groups of WUGs organized along some of the distributaries clearly show the improvements in the irrigation system, and on farms, that can follow effective organization. None of the other schemes attains this level of performance. Elsewhere at Lam Pao, the WUGs accomplish their basic purposes—to keep the tertiary canals and watercourses open and to assemble labour to help the agency keep the larger canals clear—but do little else. In Myanmar, the WUGs are subordinate to the village councils and do not seek or achieve any higher purpose. In Vietnam they are barely more than arms of the provincial irrigation authority.

Strong water user groups are not a primary cause of the relatively successful O&M observed in the schemes studied. Groups with broad participation and strong leadership enhance the efficiency of water distribution and use. But weak groups do not condemn schemes to utter inefficiency. Farmers cooperate to achieve at least basic O&M goals regardless of the maturity of the formal organization.

The experience shows there are other viable organizational models for allocating water than formal irrigator groups. The field studies identified a wide range of organizational procedures, including some reasonably well administered systems developed locally, or with targeted TA from donors, that combine hierarchical authority with user participation and are shaped by country traditions. At Dau Tieng and the left main canal at Kinda, for example, public agencies ration water intelligently according to availability, and farmers cooperate, inside or outside formal associations, to keep channels open.

Relationships between tertiary units are more problematic than those within. At this level, neighbourhood cohesion, and the other social forces that ensure reasonable cooperation within each tertiary system, are too weak to guarantee equitable water sharing. Here associations and formal federations of primary WUGs can make a substantial difference. (At Lam Pao, as the associations of WUGs sharing the same secondary canals gain strength, the functions and prominence of the watercourse WUGs themselves tend to diminish. This is predictable, because once the association of WUG leaders has determined an appropriate water-sharing formula, or cleaning schedule, meetings at the lower level can be dispensed with.)

Contrast with Flood Control

The flood control schemes in Bangladesh present a different profile, more in keeping with the paradigm of poor maintenance and lack of cooperation among farmers. Actual benefits are closer to expectations in these schemes than in the irrigation projects, but maintenance standards are deplorable and the sustainability of the structures and benefits is in doubt.

Professional incentives at the Bangladesh Water Development Board favour civil engineering skills over water management skills, and the involvement of agriculture and rural development agencies is minimal. The board has only recently made efforts to organize farmer groups for maintenance, and farmers have not associated spontaneously. Though they would all benefit from proper maintenance of older schemes, the benefits of cooperation are less certain, being reaped only when floods appear, and they are not equally shared. The potential implications for the economic and social sustainability of the major flood control investments taking place in Bangladesh are disturbing.

Looking to the Future

The turning of the terms of trade against paddy growers in the early 1980s casts a shadow over plans for better O&M. At Lam Pao, farmers who let their land lie fallow in the dry season, instead of double cropping, do not join the working groups. In Thailand and increasingly in Vietnam, families who continue to farm must contend with husbands and young adults leaving the fields and wives, and older members taking up the slack. Such trends imply a shift back to subsistence production, and less interest in or ability to do good operation and maintenance. Dry season production will be the most seriously affected by these changes, but the monsoon labour profile is also changing and with it attitudes to operation and maintenance.

Conclusions and Suggestions

Though the sample is small, the similarity of the findings across the different schemes suggests that the following lessons may have wider application:

- Tailor the prescriptions of programs for improving O&M. Agencies and irrigators do well at some functions and fail at others, often a reflection of the incentives they face. For example, exhortations to farmers to keep tertiary gates in working order, and thus risk curtailing their own water supplies, are unlikely to succeed. Hence the need to identify poorly performing components, provide incentives to bring them to appropriate standards, and tailor prescriptions based on intensive consultation with farmers and officials.

- Simplify technology. Sophisticated water distribution and monitoring technologies should be put aside in favour of controls that need less human intervention, at least until intensive diversified cropping systems are in place.

- In Bank projects, emphasize capacity building for effective water distribution associations, giving priority to federating user groups beyond the tertiary watercourse level. Hybrid organizational arrangements that take careful account of existing social networks, and that combine community labour with official agency support, should be piloted to improve the maintenance of canals and gates. Flood control and drainage schemes, where the development of user groups has been ignored, are prime candidates.

- Ensure that project engineering takes adequate account of hydrological, topographical, and social factors. If farmers are to take over responsibility for financing and developing tertiary networks, or managing irrigation, they should be involved early. Even if not formally organized, they should be brought into the design process of the irrigation system and then persuaded to enter agreements for partial financing, approval of designs, participation in construction, and management after construction is completed. Participatory project design is important everywhere but should be mandatory in flood-prone, poorly drained, densely populated areas.

- In government policy, favour crop diversification and intensification, supported by enhanced extension and marketing services. Exhortations to recover costs from farmers should be muted until water systems are reliable, more remunerative crops are introduced, and volumetric water delivery becomes practicable.

History of Irrigation Development in India

The history of irrigation development in India can be traced back to prehistoric times. Vedas and ancient Indian scriptures made references to wells, canals, tanks and dams which were beneficial to the community and their efficient operation and maintenance was the responsibility of the State. Civilization flourished on the banks of the rivers and harnessed the water for sustenance of life. According to the ancient Indian writers, the digging of a tank or well was amongst the greatest of the meritorious acts of a man. Brihaspathi, an ancient writer on law and politics, states that the construction and the repair of dams is a pious work and its burden should fall on the shoulders of rich men of the land. Vishnu Purana enjoins merit to a person who effects repairs to wells, gardens and dams.

In a monsoon climate and an agrarian economy like India, irrigation has played a major role in the production process. There is evidence of the practice of irrigation since the establishment of settled agriculture during the Indus Valley Civilization (2500 BC). These irrigation technologies were in the form of small and minor works, which could be operated by small households to irrigate small patches of land and did not require co-operative effort. Nearly all these irrigation technologies still exist in India with little technological change, and continue to be used by independent households for small holdings. The lack of evidence of large irrigation works at this time signifies the absence of large surplus that could be invested in bigger schemes or, in other words, the absence of rigid and unequal property rights. While village communities and co-operation in agriculture did exist as seen in well developed townships and economy, such co-operation in the large irrigation works was not needed, as these settlements were on the fertile and well irrigated Indus basin. The spread of agricultural settlements to less fertile and irrigated area led to co-operation in irrigation development and the emergence of larger irrigation works in the form of reservoirs and small canals. While the construction of small schemes was well within the capability of village communities, large irrigation works were to emerge only with the growth of states, empires and the intervention of the rulers. There used to emerge a close link between irrigation and the state. The king had at his disposal the power to mobilize labour which could be used for irrigation works.

In the south, perennial irrigation may have begun with construction of the Grand Anicut by the Cholas as early as second century to provide irrigation from the Cauvery river. Wherever the topography and terrain permitted, it was an old practice in the region to impound the surface drainage water in tanks or reservoirs by throwing across an earthen dam with a surplus weir, where necessary, to take off excess water, and a sluice at a suitable level to irrigate the land below. Some of the tanks got supplemental supply from stream and river channels. The entire land-scape in the central and southern India is studded with numerous irrigation tanks which have been traced back to many centuries before the beginning of the Christian era. In northern India also there are a number of small canals in the upper valleys of rivers which are very old.

Soil and Water Relationships

Soil moisture limits forage production potential the most in semiarid regions. Estimated water use efficiency for irrigated and dryland crop production systems is 50 percent, and available soil water has a large impact on management decisions producers make throughout the year. Soil moisture available for plant growth makes up approximately 0.01 percent of the world's stored water. By understanding a little about the soil's physical properties and its relationship to soil moisture, you can make better soil-management decisions. Soil texture and structure greatly influence water infiltration, permeability, and water-holding capacity.

Soil texture refers to the composition of the soil in terms of the proportion of small, medium, and large particles (clay, silt, and sand, respectively) in a specific soil mass. For example, a coarse soil is a sand or loamy sand, a medium soil is a loam, silt loam, or silt, and a fine soil is a sandy clay, silty clay, or clay.

Soil structure refers to the arrangement of soil particles (sand, silt, and clay) into stable units called aggregates, which give soil its structure. Aggregates can be loose and friable, or they can form distinct, uniform patterns. For example, granular structure is loose and friable, blocky structure is six-sided and can have angled or rounded sides, and platelike structure is layered and may indicate compaction problems.

Soil porosity refers to the space between soil particles, which consists of various amounts of water and air. Porosity depends on both soil texture and structure. For example, a fine soil has smaller but more numerous pores than a coarse soil. A coarse soil has bigger particles than a fine soil, but it has less porosity, or overall pore space. Water can be held tighter in small pores than in large ones, so fine soils can hold more water than coarse soils.

Water infiltration is the movement of water from the soil surface into the soil profile. Soil texture, soil structure, and slope have the largest impact on infiltration rate. Water moves by gravity into the open pore spaces in the soil, and the size of the soil particles and their spacing determines how much water can flow in. Wide pore spacing at the soil surface increases the rate of water infiltration, so coarse soils have a higher infiltration rate than fine soils.

Permeability refers to the movement of air and water through the soil, which is important because it affects the supply of root-zone air, moisture, and nutrients available for plant uptake. A soil's permeability is determined by the relative rate of moisture and air movement through the most restrictive layer within the upper 40 inches of the effective root zone. Water and air rapidly permeate coarse soils with granular subsoils, which tend to be loose when moist and don't restrict water or air movement. Slow permeability is characteristic of a moderately fine subsoil with angular to subangular blocky structure. It is firm when moist and hard when dry.

Water-holding capacity is controlled primarily by soil texture and organic matter. Soils with smaller particles (silt and clay) have a larger surface area than those with larger sand particles, and a large surface area allows a soil to hold more water. In other words, a soil with a high percentage of silt and clay particles, which describes fine soil, has a higher water-holding capacity. The table illustrates water-holding-capacity differences as influenced by texture. Organic matter percentage also influences water-holding capacity. As the percentage increases, the water-holding capacity increases because of the affinity organic matter has for water.

Water availability is illustrated in the figure by water levels in three different soil types. Excess or gravitational water drains quickly from the soil after a heavy rain because of gravitational forces (saturation point to field capacity). Plants may use small amounts of this water before it moves out of the root zone. Available water is retained in the soil after the excess has drained (field capacity to wilting point). This water is the most important for crop or forage production. Plants can use approximately 50 percent of it without exhibiting stress, but if less than 50 percent is available, drought stress can result. Unavailable water is soil moisture that is held so tightly by the soil that it cannot be extracted by the plant. Water remains in the soil even below plants' wilting point.

Soil is a valuable resource that supports plant life, and water is an essential component of this system. Management decisions concerning types of crops to plant, plant populations, irrigation scheduling, and the amount of nitrogen fertilizer to apply depend on the amount of moisture that is available to the crop throughout the growing season. By understanding some physical characteristics of the soil, you can better define the strengths and weaknesses of different soil types.

Soil, Water and Plant Characteristics Important to Irrigation

Irrigation, applying water to assure sufficient soil moisture is available for good plant growth, as practiced in North Dakota is called "supplemental irrigation" because it is used to augment the rainfall that occurs during the growing season. Irrigation is used on full season agronomic crops to provide a dependable yield every year. It is also used on crops where water stress affects the quality of the yield, such as flowers, vegetables and fruits.

During most years it is not uncommon for some places in the state to receive sufficient rainfall for good plant growth while other areas experience reduced yields or quality on non-irrigated crops because of water stress from insufficient soil moisture. For irrigation planning purposes, average precipitation during the growing season is not a good yardstick for determining a need for irrigation. The timing and amounts of rainfall during the season, the soil's ability to hold water, and the crop's water requirements are all factors which influence the need for irrigation. Any location in the state can have what might be considered "wet" or "dry" weeks, months and even years. Under irrigation, soil and water compatibility is very important. If they are not compatible, the applied irrigation water could have an adverse effect on the chemical and physical properties of the soil. Determining the suitability of land for irrigation requires a thorough evaluation of the soil properties, the topography of the land within the field and the quality of water to be used for irrigation. A basic understanding of soil/water/plant interactions will help irrigators efficiently manage their crops, soils, irrigation systems and water supplies.

Soil Properties

Soil surveys of every county in North Dakota have been completed by the Natural Resource Conservation Service (NRCS, formerly the SCS). The county soil survey report provides detailed soils information on any parcel of land and is available from the county NRCS office or the NDSU Department of Soil Science. The soil properties of texture, structure, depth, permeability and chemistry play an important role in irrigation management.

Soil Texture

Soil texture is determined by the size and type of solid particles that make up the soil. Soil particles may be either mineral or organic. In most soils, the largest proportion of particles are mineral and are referred to as "mineral soils." For mineral soils, the texture is based on the relative proportion of the particles under 2 millimetres (mm) or 5/64th of an inch in size. Soil texture classes may be modified if greater than 15% of the particles are organic (e.g. mucky silt loam). Soil particles greater than 2 mm in size are not used to determine soil texture. However, when they make up more than 15% of the soil volume, the textural class is modified (e.g. gravelly sand).

Soil texture can be determined by separating and weighing the sand, silt and clay. For example, if a 100 pound sample of soil was sifted through screens and found to contain 45 pounds of sand, 35 pounds of silt and 20 pounds of clay, then the soil would be composed of 45% sand, 35% silt and 20% clay. As shown by the dotted lines, this soil has a loam texture. There are 12 basic soil textures. Sand, loamy sands and sandy loams are the most common soil textures irrigated in North Dakota.

Soil Structure

Soil structure refers to the grouping of particles of sand, silt, and clay into larger aggregates of various sizes and shapes. The processes of root penetration, wetting and drying cycles, freezing and thawing, and animal activity combined with inorganic and organic cementing agents produce soil structure. Structural aggregates that are resistant to physical stress are important to the maintenance of soil tilth and productivity. Practices such as excessive cultivation or tillage of wet soils disrupt aggregates and accelerate the loss of organic matter, causing decreased aggregate stability.

The movement of air, water, and plant roots through a soil is affected by soil structure. Stable aggregates result in a network of soil pores that allow rapid exchange of air and water with plant roots. Plant growth depends on rapid rates of exchange. Good soil structure can be maintained by practicing beneficial soil management such as crop rotations, organic matter additions, and timely tillage practices. In sandy soils, aggregate stability is often difficult to maintain due to low organic matter, clay content and resistance of sand particles to cementing processes.

Soil Series

Soil is the layer of the earth's surface which has been changed by physical or biological processes. The five soil-forming factors that control the process of change are parent material, climate, topography, biota (plants and animals) and time. Soils are grouped into categories according to their observed properties. The USDA classification system consists of six categories. The highest category (soil order) contains 11 basic soil groups, each with a very broad range of properties. The lowest category (soil series) contains over 12,000 soils, each defining a very narrow range in soil properties.

North Dakota has 264 soil series. A soil series is unique because of a combination of properties such as texture, structure, topographic position (on the side of a hill or in a valley) or depth to the water table. A particular soil series describes locations where these soil conditions are similar.

These locations may be in the same field, section, county, state or even region. Soil delineations on county soil survey maps are based on the soil series. A soil series is generally named after a town near the site that represents the typical properties for that soil. For example, the site with typical properties for the Embden soil series is near Embden, North Dakota.

Many soil series do not have a deep, uniform soil profile. Restrictive subsurface layers often interfere with root penetration. In these situations the roots will be concentrated in the upper part of the soil profile. For example, in the Renshaw loam profile, the majority of the plant roots will be in the top 18 inches because of the poor growing environment encountered in the underlying sand and gravel substrata. This type of information is important for irrigation management.

Soil Depth

Soil depth refers to the thickness of the soil materials which provide structural support, nutrients, and water for plants. In North Dakota, soil series that have bedrock between 10 and 20 inches from the surface are described as shallow. Bedrock between 20 and 40 inches is described as moderately deep. Most soil series in North Dakota have bedrock at depths greater than 40 inches and are described as deep. Depth to contrasting textures is given in the soil series descriptions in the county soil survey report. The depth to a contrasting soil layer of sand and gravel can affect irrigation management decisions. If the depth to this layer is less than 3 feet, the rooting depth and available soil water for plants is decreased. Soils with less available water for plants require more frequent irrigations.

Soil Permeability and Infiltration

A soil's permeability is a measure of the ability of air and water to move through it. Permeability is influenced by the size, shape, and continuity of the pore spaces, which in turn are dependent on the soil bulk density, structure and texture. Most soil series are assigned to a single permeability class based on the most restrictive layer in the upper 5 feet of the soil profile. However, soil series with contrasting textures in the soil profile are assigned to more than one permeability class. In most cases, soils with a slow, very slow, rapid or very rapid permeability classification are considered poor for irrigation.

Soil Permeability Classes

Classification	Infiltration Rate inches/hour)
Very Slow	Less than 0.06
Slow	0.06 to 0.2
Moderately Slow	0.2 to 0.6
Moderate	0.6 to 2.0
Moderately Rapid	2.0 to 6.0
Rapid	6.0 to 20.0
Very Rapid	Greater than 20.0

Infiltration is the downward flow of water from the surface through the soil. The *infiltration rate* (sometimes called intake rate) of a soil is a measure of its ability to absorb an amount of rain or irrigation water over a given time period. It is commonly expressed in inches per hour. It is dependent on the permeability of the surface soil, moisture content of the soil and surface conditions such as roughness (tillage and plant residue), slope, and plant cover.

Coarse textured soils such as sands and gravel usually have high infiltration rates. The infiltration rates of medium and fine textured soils such as loams, silts, and clays are lower than those of coarse textured soils and more dependant on the stability of the soil aggregates. Water and plant nutrient losses may be greater on coarse textured soils, so the timing and quantity of chemical and water applications is particularly critical on these soils.

Saline and Sodic Soils

Salt affected soils are grouped according to their content of soluble salts and sodium. Saline and sodic soils usually occur in areas where ground water moves upward from a shallow water table close to the soil surface. The water carries salts which accumulate in the soil as the water is evaporated from the soil surface or transpired through the plants to the atmosphere.

In General, these Soils are not Recommended for Irrigation

Saline and sodic soils may be of natural or man-made origins. One of the man-made processes is related to irrigation. Under certain combinations of irrigation water quality and soils, salts and/or sodium may accumulate in the root zone and have an adverse effect on plant growth. Under some conditions, *sodium* can be controlled in the upper part of the soil through the use of calcium amendments. The replacement of sodium by calcium improves the structure of the soil. Calcium soil amendments can be helpful in situations where land with a majority of unaffected irrigable soils contains pockets (inclusions) of sodium affected soils. Under irrigation, calcium soil amendments will help where surface crusting has become a problem. Special irrigation management practices may be required on these soils.

Salt concentrations can be managed by leaching or controlling the water table elevation. Leaching is accomplished by applying more water than the soil will hold within the root zone. Large rainfall events, applying additional irrigation water or both will carry some of the salts below the root zone. Water table control can be accomplished by planting a deep rooted crop, such as alfalfa, or installing subsurface drainage. Deep ditches and tiling are methods of subsurface drainage that have been used successfully to control the level of the water table in many parts of the world.

Soil salt and sodium contents need to be measured to precisely determine the severity of the problem. The salt content of the soil is estimated from an electrical conductivity measurement using a soil water extract, soil water slurry or soil paste. The sodium content of the soil is often measured on a soil water extract and expressed as the ratio between the sodium and calcium plus magnesium and given the term sodium adsorption ratio (SAR).

Soils can be monitored by soil sampling the surface layer (top 6 inches) on a periodic basis (every three to five years). The SAR of the soil samples will indicate if there is a buildup of sodium. Generally, soils with an SAR of 13 from the saturated extract will exhibit significant physical problems due to dispersal of clay particles. Usually a soil with an SAR of 6 or lower from the saturated extract will not have physical problems associated with dispersed clay. However, if periodic sampling indicates that the SAR is increasing, say from 6 to 9, then it may be time to consider corrective action.

Topography of the Field

Topography or the "lay of the land" has a large impact on whether a field can be irrigated. Relief is a component of topography that refers to the difference in height between the hills and depressions in the field. The topographic relief will affect the type of irrigation system to be used, the water conveyance system (ditches or pipes), drainage requirements and water erosion control practices. The shape and arrangement of topographic landforms and the type of surface waterway network will also influence irrigation management.

Slope

Slope is important to soil formation and management because of its influence on runoff, soil drainage, erosion, use of machinery, and choice of crops. Slope is the incline or gradient of a surface and is commonly expressed in percent. The percent slope is determined by measuring the difference in vertical elevation in feet over 100 feet of horizontal distance. For example, a 5 percent slope rises or falls 5 feet per 100 feet of horizontal distance. In addition to the percent of slope, the shape of the slope is another important characteristic. A convex slope curves outward like the outside surface of a ball, a concave slope curves inward like the inside surface of a saucer, and a plane slope is like a tilted flat surface. Slopes are described as simple or complex. Simple slopes have a smooth appearance with surfaces extending in one or perhaps two directions. For example, slopes on alluvial fans and foot slopes of river valleys are regarded as simple. Complex areas have short slopes which extend in several directions and consist of convex and concave slopes much like the knoll and pothole topography found on glacial till plains.

Simple slopes of 1% or less are commonly used for gravity (surface) irrigation. Simple and complex slopes greater than 1% should only be irrigated with sprinkler or drip systems. Centre pivot sprinkler irrigation systems can operate on slopes up to 15%, but simple slopes greater than 9% are not generally recommended. To accommodate an irrigation application method such as gravity or sprinkler systems, the slope in a field can be modified by land smoothing. However, land smoothing may cause yield reductions for one to three growing seasons. The places where topsoil was removed are most likely to have yield reductions. Special management of these areas through increased fertilizer and organic matter applications may be required for accelerated recovery.

Irrigation Water Quality

The quality of some water is not suitable for irrigating crops. Irrigation water must be compatible with both the crops and soils to which it will be applied. The Soil and Water Environmental Laboratory in the NDSU soil science department provides soil and water compatibility recommendations for irrigation. Generally a water analysis and a legal description of the land proposed for irrigation are required before a recommendation can be made. The quality of water for irrigation purposes is determined by its salt content. An analysis of water for irrigation should include the *cations*: calcium, magnesium, and sodium, and the *anions*: bicarbonate, carbonate, sulfate, and chloride. Some crops are sensitive to boron, so it is often included in the analysis.

Irrigation Water Classification

The two most important factors to look for in an irrigation water quality analysis are the Total Dissolved Solids (TDS) and the Sodium Adsorption Ratio (SAR). The TDS of a water sample is a measure of the concentration of soluble salts in a water sample and is commonly referred to as the *salinity* of the water. TDS is expressed in terms of the electrical conductivity (EC) and its units are either:

- millimhos per centimetre (mmhos/cm),
- deci-Siemens per meter (dS/m) or
- micromhos per centimetre (mmhos/cm).

Where:

- 1000 mmhos/cm = 1 mmho/cm = 1 dS/m.

The SAR of a water sample is the proportion of sodium relative to calcium and magnesium. Since it is a ratio, the SAR has no units.

Laboratories that perform irrigation water analysis may provide a suitability classification based on a system developed at the U.S. Salinity Laboratory in California. This classification system combines salinity and sodicity. For example, a water sample classified as C3-S2 would have a high salinity rating and a medium sodium rating. The scale for sodicity is not constant because it depends on the level of salinity. For example, an SAR of 8 is in the S1 category if the salinity is from 100 to 300 mmhos/cm; S2 if the salinity is from 300 to 3000 mmhos/cm, and S3 if the salinity is greater than 3000 mmhos/cm.

Much of the water in North Dakota is classified in the C2 to C3 salinity range and the S1 to S2 sodium hazard range. In general, any water with an EC greater than 2000 mmhos/cm or an SAR value greater than 6 is not recommended for continuous irrigation in North Dakota. In cases where sporadic irrigation is practiced (i.e. a particular piece of land is only irrigated one year out of three or more), lower quality water may be used. However, the lower quality water should not have an EC that exceeds 3000 mmhos/cm or an SAR greater than 10.

Calcium added to irrigation water can lower the SAR and reduce the harmful effects of sodium. The effectiveness of added calcium depends on its solubility in the irrigation water.

Calcium solubility is controlled by both the source of the calcium (e.g. calcium carbonate, gypsum, calcium chloride) and also the concentration of other ions in the irrigation water. Compared to calcium carbonate and gypsum, calcium chloride additions will result in higher concentrations of soluble calcium and be the most effective at lowering irrigation water SAR. However, calcium chloride is considerably more expensive than calcium carbonate and calcium sulfate (gypsum).

Carbonates

Carbonate and bicarbonate ions in the water combine with calcium and magnesium to form compounds which precipitate out of solution. Removing calcium and magnesium increases the sodium hazard to the soil from irrigation water. The increased sodium hazard is often expressed as "adjusted SAR." The increase of "adjusted SAR" over the SAR is a relative indication of the increase in sodium hazard due to the presence of these ions. Precipitation of carbonate minerals has not been observed to plug sprinkler systems in North Dakota, but these minerals can cause plugging in drip irrigation systems. To control this problem, the pH of the irrigation water is generally lowered by adding a mild acid.

Salinity

C1-Low Salinity Water: can be used for irrigation with most crops on most soils with little likelihood that soil salinity will develop. Some leaching is required, but this occurs under normal irrigation practices except in soils of slow and very slow permeability.

C2-Medium Salinity Water: can be used if a moderate amount of leaching occurs. In most cases plants with moderate salt tolerance can be grown without special practices for salinity control.

C3-High Salinity Water: cannot be used on soils with moderately slow to very slow permeability. Even with adequate permeability, special management for salinity control may be required and plants with good salt tolerance should be selected.

C4-Very High Salinity Water: is not suitable for irrigation under ordinary conditions, but may be used occasionally under very special circumstances. The soils must have rapid permeability, drainage must be adequate, irrigation water must be applied in excess to provide considerable leaching, and very salt tolerant crops should be selected.

Sodium

S1-Low Sodium Water: can be used for irrigation on almost all soils with little danger of development of harmful levels of exchangeable sodium.

S2-Medium Sodium Water: will present an appreciable sodium hazard in fine textured soils, especially under low leaching conditions. This water may be used on coarse textured soils with moderately rapid to very rapid permeability.

S3-High Sodium Water: will produce harmful levels of exchangeable sodium in most soils and requires special soil management, good drainage, high leaching, and high organic matter additions.

S4-Very High Sodium Water: is generally unsatisfactory for irrigation purposes except at low and perhaps medium salinity.

Boron

Boron is essential for the normal growth of all plants, but the quantity required is very small. Plants sensitive to boron, such as dry beans, require much smaller amounts than plants that are tolerant of boron, such as corn, potatoes and alfalfa. In fact, the concentration of boron that will injure the sensitive plants is often close to that required for normal growth of tolerant plants.

Although there have been no documented problems with boron in water used for irrigation in North Dakota, testing for this element in irrigation water is a precautionary practice. Boron does occur in some North Dakota ground water at concentrations that are theoretically toxic to some crops. Boron concentration greater than 2 parts per million (ppm) may be a problem for certain sensitive crops, especially in years that require large quantities of irrigation water.

The Interaction Between Soil and Water

Soil is a medium that stores and moves water. If a cubic foot of a typical silt loam topsoil were separated into its component parts, about 45% of the volume would be mineral matter (soil particles), organic residue would occupy about 5% of the volume, and the rest would be *pore space*. The pore space is the voids between soil particles and is occupied by either air or water. The quantity and size of the pore spaces are determined by the soil's texture, bulk density and structure.

Water is held in soil in two ways: as a thin coating on the outside of soil particles and in the pore spaces. Soil water in the pore spaces can be divided into two different forms: gravitational water and capillary water. Gravitational water generally moves quickly downward in the soil due to the force of gravity. Capillary water is the most important for crop production because it is held by soil particles against the force of gravity.

As water infiltrates into a soil, the pore spaces fill with water. As the pores are filled, water moves through the soil by gravity and capillary forces. Water movement continues downward until a balance is reached between the capillary forces and the force of gravity. Water is pulled around soil particles and through small pore spaces in any direction by capillary forces. When capillary forces move water from a shallow water table upward, salts may precipitate and concentrate in the soil as water is removed by plants and evaporation.

Water Holding Capacity of Soils

There are four important levels of soil moisture content that reflect the availability of water in the soil. These levels are commonly referred to as: 1) saturation, 2) field capacity, 3) wilting point and 4) oven dry. When a soil is saturated, the soil pores are filled with water and nearly all of the air in the soil has been displaced by water. The water held in the soil between saturation and field capacity is gravitational water. Frequently, gravitational water will take a few days to drain through the soil profile and some can be absorbed by roots of plants.

Field capacity is defined as the level of soil moisture left in the soil after drainage of the gravitational water. Water held between field capacity and the wilting point is available for plant use. The wilting point is defined as the soil moisture content where most plants cannot exert enough force to remove water from small pores in the soil. Most crops will be permanently damaged if the soil moisture content is allowed to reach the wilting point. In many cases, yield reductions may occur long before this point is reached. Capillary water held in the soil beyond the wilting point can only be removed by evaporation. When soil is dried in an oven, nearly all water is removed. "Oven dry" moisture content is used to provide a reference for measuring the other three soil moisture contents.

When discussing the water holding capacity associated with a particular soil series, the water available for plant use in the *root zone* is commonly given. Available soil water content is commonly expressed as inches per foot of soil. For example, the water available can be calculated for a soil with fine sandy loam in the first foot, loamy sand in the second foot and sand in the third foot. The top foot would have about 2.0 inches, the second foot would have about 1.0 inch and the third foot would have about 0.75 inches for a total of 3.75 inches of available water for a crop with a 3 foot root depth.

Soil Moisture Tension

The degree to which water clings to the soil is the most important soil water characteristic to a growing plant. This concept is often expressed as soil moisture tension. Soil moisture tension is negative pressure and commonly expressed in units of *bars*. During this discussion, when soil moisture tension becomes more negative it will be referred to as "increasing" in value.

Thus, as soil moisture tension increases (the soil water pressure becomes more negative), the amount of energy exerted by a plant to remove the water from the soil must also increase. One bar of soil moisture tension is nearly equivalent to-1 atmosphere of pressure (1 atmosphere of pressure is equal to 14.7 pounds per square inch at sea level). A soil that is saturated has a soil moisture tension of about 0.001 bars, or less, which requires little energy for a plant to pull water away from the soil. At field capacity most soils have a soil moisture tension between 0.05 and 0.33 bars. Soils classified as sandy may have field capacity tensions around 0.10 bars, while clayey soil will have field capacity at a tension around 0.33 bars. At field capacity it is relatively easy for a plant to remove water from the soil.

The wilting point is reached when the maximum energy exerted by a plant is equal to the tension with which the soil holds the water. For most agronomic crops this is about 15 bars of soil moisture tension. To put this in perspective, the wilting point of some desert plants has been measured between 50 and 60 bars of soil moisture tension. The presence of high amounts of soluble salts in the soil reduces the amount of water available to plants. As salts increase in soil water, the energy expended by a plant to extract water must also increase, even though the soil moisture tension remains the same. In essence, salts decrease the total available water in the soil profile.

How Plants get Water from Soil

Water is essential for plant growth. Without enough water, normal plant functions are disturbed, and the plant gradually wilts, stops growing, and dies. Plants are most susceptible to damage from water deficiency during the vegetative and reproductive stages of growth. Also, many plants are most sensitive to salinity during the germination and seedling growth stages. Most of the water that enters the plant roots does not stay in the plant. Less than 1% of the water withdrawn by the plant is actually used in photosynthesis (i.e. assimilated by the plant). The rest of the water moves to the leaf surfaces where it transpires (evaporates) to the atmosphere. The rate at which a plant takes up water is controlled by its physical characteristics, the atmosphere and soil environment. As water moves from the soil, into the roots, through the stem, into the leaves and through the leaf stomata to the air, it moves from a low water tension to a high water tension. The water tension in the air is related to its relative humidity and is always greater than the water tension in the soil.

Plants can extract only the soil water that is in contact with their roots. For most agronomic crops, the root distribution in a deep uniform soil is concentrated near the soil surface. Over the course of a growing season, plants generally extract more water from the upper part of their root zone than from the lower part.

Plants such as grasses, with a high root density per unit of soil volume, may be able to absorb all available soil water. Other plants, such as vegetables, with a low root density, may not be able to obtain as much water from an equal volume of the same soil. Vegetables are generally more sensitive to water stress than high root density agronomic crops such as alfalfa, corn, wheat and sunflower.

Crop Water Use

Crop water use, also called evapotranspiration or ET, is an estimate of the amount of water transpired by the plants and the amount of evaporation from the soil surface around the plants. A plant's water use changes with a predictable pattern from germination to maturity. All agronomic crops have a similar water use pattern. However, crop water use can change from growing season to growing season due to changes in climatic variables (air temperature, amount of sunlight, humidity, wind) and soil differences between fields (root depth, soil water holding capacities, texture, structure, etc.). Many years of research have produced a number of equations that allow accurate estimates of crop water use values to be calculated from measured daily weather variables. Accurate estimates of crop water use values can be calculated for all the major irrigated crops in North Dakota.

Knowledge of water use patterns during the different growth stages has a major influence on how an irrigation system is designed and managed. Failure to recognize the water use patterns of a crop may result in poorly managed water applications. Crop water stress, fertilizer and pesticide leaching and increased pumping costs are just a few of the results of poor irrigation water management.

Irrigation Water Management

Obtaining increased yield from irrigation requires appropriate management of all the inputs. This means fertilizing to meet the yield goal, good tillage practices and efficient management of the amount of applied water. One of the most difficult parts of irrigation management is deciding when to turn on the irrigation system and how much water to apply. Fortunately, irrigation scheduling methods to help make those decisions have been developed.

Using rational or scientific methods to schedule irrigations is essential for good irrigation management, especially in North Dakota where irrigation is used to supplement rain. Good irrigation management begins with accurate measurement of the rain received on each irrigated field and knowing the soil moisture status in each field at the start of the vegetative growth stage. Over the years, a number of scheduling methods have been developed. Measurement of soil moisture levels has been the most common method of irrigation scheduling, but newer methods use a combination of crop water use and soil water estimates.

The oldest and most commonly used irrigation scheduling method is the "feel method," which estimates soil moisture by taking a soil sample in hand and squeezing it into a ball, observing the appearance of the ball and creating a ribbon of soil between the thumb and forefinger to estimate the soil moisture content. This method requires practice and experience to become accurate at predicting irrigation water needs. It is popular because it can be combined with other field activities such as scouting for insects, soil sampling for nitrogen, petiole sampling, etc. More accurate soil moisture measurement methods use mechanical devices such as tensiometers and soil moisture blocks for irrigation scheduling. These devices are particularly helpful with fruit and vegetable crops and have proven to be accurate, reliable and inexpensive. Other more sophisticated instrumentation can be used for irrigation scheduling but generally are not used for irrigation management because of the expense. An irrigation scheduling procedure called the "checkbook" method has been used successfully for many years in North Dakota. The checkbook method is a soil moisture accounting method which uses crop water use values and soil water holding capacities to predict the time to irrigate and amount of water needed to replenish what has been removed from the root zone.

North Dakota has a number of automated weather stations which record weather data on an hourly basis. This system is called the North Dakota Agricultural Weather Network (NDAWN). The weather data collected at each station allows calculation of accurate estimates of crop water use values on a daily basis. The crop water use estimates for several crops are available electronically (bulletin board) for each weather station on the NDAWN system and can be used with the checkbook method. This new technology now provides a way to access site-specific estimates of crop water use values.

What is Soil Moisture?

Soil moisture is difficult to define because it means different things in different disciplines. For example, a farmer's concept of soil moisture is different from that of a water resource manager or a weather forecaster. Generally, however, soil moisture is the water that is held in the spaces between soil particles. Surface soil moisture is the water that is in the upper 10 cm of soil, whereas root zone soil moisture is the water that is available to plants, which is generally considered to be in the upper 200 cm of soil.

Why is Measuring Soil Moisture Important?

Compared to other components of the hydrologic cycle, the volume of soil moisture is small; nonetheless, it of fundamental importance to many hydrological, biological and biogeochemical processes. Soil moisture information is valuable to a wide range of government agencies and private companies concerned with weather and climate, runoff potential and flood control, soil erosion and slope failure, reservoir management, geotechnical engineering, and water quality. Soil moisture is a key variable in controlling the exchange of water and heat energy between the land surface and the atmosphere through evaporation and plant transpiration. As a result, soil moisture plays an important role in the development of weather patterns and the production of precipitation. Simulations with numerical weather prediction models have shown that improved characterization of surface soil moisture, vegetation, and temperature can lead to significant forecast improvements. Soil moisture also strongly affects the amount of precipitation that runs off into nearby streams and rivers. Large-scale dry or wet surface regions have been observed to impart positive feedback on subsequent precipitation patterns, such as in the extreme conditions over the central U.S. during the 1988 drought and the 1993 floods. Soil moisture information can be used for reservoir management, early warning of droughts, irrigation scheduling, and crop yield forecasting.

Remote Sensing of Soil Moisture

Despite the importance of soil moisture information, widespread and/or continuous measurement of soil moisture is all but nonexistent. "The lack of a convincing approach of global measurement of soil moisture is a serious problem" (National Research Council, 1992). Clearly, a need exists for continuous measurements of surface soil moisture with global coverage. Remote sensing of soil moisture from the vantage point of space is advantageous because of its spatial coverage and temporal continuity, but this capability does not yet exist. Research in soil moisture remote sensing began in the mid 1970's shortly after the surge in satellite development. Subsequent research has occurred along many diverse paths. Quantitative measurements of soil moisture in the surface layer of soil have been most successful using passive remote sensing in the microwave region. The potential exists today to retrieve soil moisture estimates from space-based instruments at frequencies of about 6 GHz (C band). However, observations at frequencies between 1 and 3 GHz (L band) are best suited for detection of soil moisture because energy is emitted from a deeper soil layer and less energy is absorbed or reflected by vegetation.

Soil moisture remote sensing is fraught with challenges. Only the moisture in the top few centimeters of soil can be detected. Algorithm development is complicated by the need for surface roughness and vegetation corrections, which are based on empirical relationships of limited breadth. Extending ground-based techniques to space-based systems requires innovative antenna technology. In spite of these challenges, recent advances in aperture synthesis and thinned array technology applied at L band have shown great promise for soil moisture mapping. Scientists that the Global Hydrology and Climate Centre have been involved in experiments to address some the challenges we face in developing this technology.

Defining the Range of Uncertainty Associated with Remotely Sensed Soil Moisture

As science and technology continue to advance towards establishing a capability to remotely sense soil moisture, we must periodically evaluate our progress. From such introspection, we gain insights into our strengths and weakness and can identify research foci. Retrieval of soil moisture information by remote sensing requires, among other things, soil property information and corrections for surface roughness and vegetation optical depth. Current correction procedures are based on empirical relationships of limited breadth. Extending ground-based techniques to space-based systems requires using remote sensing to provide estimates of the input variables and parameters. The applicability of doing this was recently demonstrated during the Southern Great Plains 1997 (SGP97) Hydrology Experiment (Jackson et al., 1999). However, the potential errors in the soil moisture estimates were not addressed.

Therefore, a study was conducted to define the extent to which errors in estimating algorithm input parameters at regional scale result in uncertainty in the true value of soil moisture. From a review of procedures used to estimate the variables and parameters required to retrieve soil moisture at regional scale, we defined the range of errors generally associated with each measurement. The model was run while varying one parameter at a time over a range of values (errors) to quantify the effects individual parameters have on soil moisture retrieval. The parameters that produced the greatest effect were varied in paired combinations to characterize potential compounded effects. We then recalculate soil moisture for the Southern Great Plains 1997 Hydrology Experiment using random combinations of errors in the input data. These results are compared with the initial calculations.

Results

Although errors in estimating most soil moisture algorithm parameters at regional scale yield total variations of less than about ±2% volumetric water content (vwc), errors in estimating vegetation water content, vegetation b parameter, percent clay, and surface roughness yield much larger uncertainty in estimated soil moisture. The effects of these parameter variations on calculated soil moisture are greater for wetter soils (above 25% vwc) and can result in total errors in soil moisture retrieval of up to ±12% vwc. Errors in estimating these same parameters have a compound effect on calculated soil moisture when they vary collectively, and may yield variations in soil moisture retrieval as high as 38% vwc (-12%, +26%) for wet soil. Even under drier conditions between 10% and 25% vwc, parameter errors result in soil moisture uncertainties of ±4.5% to ±7.5%. Such uncertainties are unacceptably high for many applications and may preclude using these data where individual pixel values are of interest.

When combinations of random errors in vegetation water content, percent clay, and surface roughness are imposed on the Southern Great Plains 1997 (SGP97) Hydrology Experiment input data set, the macrostructure is reproduced but the resulting soil moisture field is significantly more heterogeneous than the original soil moisture field. Parameter estimation in SGP97 based on field observations and a supervised land cover classification contributed to a relatively low soil moisture uncertainty ($\pm 2_\sigma$) of ±3% vwc for most brightness temperatures. The calculations with larger parameter errors resulted in a larger soil moisture uncertainty of ±5% vwc for given brightness temperatures. This represents the maximum uncertainty associated with a large number of measurements using available regional scale data to estimate algorithm input parameters.

In summary, errors in regional scale estimates of vegetation water content, the b parameter, percent clay and surface roughness may lead to very substantial errors in retrieved soil moisture. These errors are particularly high for wetter soils. Such errors may preclude using these data where individual pixel values are of interest. On a regional scale, however, where samples sizes are large, uncertainties in remotely sensed soil moisture are within acceptable limits for most disciplines.

Soil Salinity

Salt-affected soils are visible on rangeland in Colorado. Salts dissolved from the soil accumulate at the soil surface and are deposited on the ground and at the base of the fence post.

Causes of Soil Salinity

Salt affected soils are caused by excess accumulation of salts, typically most pronounced at the soil surface. Salts can be transported to the soil surface by capillary transport from a salt laden water table and then accumulate due to evaporation; they can also be concentrated in soils due to human activity. As soil salinity increases, salt effects can result in degradation of soils and vegetation.

Salinization is a process that results from:

- high levels of salt in the soils.
- landscape features that allow salts to become mobile. (movement of water table)
- climatic trends that favour accumulation.
- human activities such as land clearing and aquaculture activities.

Natural Occurrence

Salt is a natural element of soils and water. The ions responsible for salinization are: Na^+, K^+, Ca^{2+}, Mg^{2+} and Cl^-. As the Na^+ (sodium) predominates, soils can become *sodic*. Sodic soils present particular challenges because they tend to have very poor structure which limits or prevents water infiltration and drainage.

Over eons, as soil minerals weather and release salts, these salts are flushed or leached out of the soil by drainage water in areas with sufficient precipitation. In addition to mineral weathering, salts are also deposited via dust and precipitation. In dry regions salts may accumulate, leading to naturally saline soils. This is the case, for example, in large parts of Australia. Human practices can increase the salinity of soils by the addition of salts in irrigation water. Proper irrigation management can prevent salt accumulation by providing adequate drainage water to leach added salts from the soil. Disrupting drainage patterns that provide leaching can also result in salt accumulations. An example of this occurred in Egypt in 1970 when the Aswan High Dam was built. The change in the level of ground water before the construction had enabled soil erosion, which led to high concentration of salts in the water table. After the construction, the continuous high level of the water table led to the salination of the arable land.

Dry Land Salinity

Salinity in drylands can occur when the water table is between two to three metres from the surface of the soil. The salts from the groundwater are raised by capillary action to the surface of the soil. This occurs when groundwater is saline (which is true in many areas), and is favoured by land use practices allowing more rainwater to enter the aquifer than it could accommodate. For example, the clearing of trees for agriculture is a major reason for dryland salinity in some areas, since deep rooting of trees has been replaced by shallow rooting of annual crops.

Salinity Due to Irrigation

Rain or irrigation, in the absence of leaching, can bring salts to the surface by capillary action. Salinity from irrigation can occur over time wherever irrigation occurs, since almost all water (even natural rainfall) contains some dissolved salts. When the plants use the water, the salts are left behind in the soil and eventually begin to accumulate. Since soil salinity makes it more difficult for plants to absorb soil moisture, these salts must be leached out of the plant root zone by applying additional water. This water in excess of plant needs is called the leaching fraction. Salination from irrigation water is also greatly increased by poor drainage and use of saline water for irrigating agricultural crops.

Salinity in urban areas often results from the combination of irrigation and groundwater processes. Irrigation is also now common in cities (gardens and recreation areas).

Consequences of Salinity

The consequences of salinity are:

- detrimental effects on plant growth and yield
- damage to infrastructure (roads, bricks, corrosion of pipes and cables)
- reduction of water quality for users, sedimentation problems

- soil erosion ultimately, when crops are too strongly affected by the amounts of salts.

Salinity is an important land degradation problem. Soil salinity can be reduced by leaching soluble salts out of soil with excess irrigation water. High levels of soil salinity can be tolerated if salt-tolerant plants are grown. A comprehensive treatment of soil salinity is available from the FAO.

Regions Affected

From the FAO/UNESCO Soil Map of the World the following salinised areas can be derived.

Region	Area (10⁶ha)
Africa	69.5
Near and Middle East	53.1
Asia and Far East	19.5
Latin America	59.4
Australia	84.7
North America	16.0
Europe	20.7

Theoretical Alternative Energy Source

Salination is theoretically an alternative energy source. The mixing of fresh and salty water releases energy; this means that devices could be located at points where fresh water enters the ocean in order to harness the energy.

Assessment of the Impact of Industrial Effluents on the Quality of Irrigation Water and Changes on Soil Characteristics (a Case of Kombolcha Town). With the ever increasing demand on irrigation water supply, farmlands are frequently faced with utilization of poor quality irrigation water. In many parts of Ethiopia, waste water, which are disposed to wells, ponds, streams and treatment plants, are used as a source of irrigation water as well as for drinking (Alemtsehaye, 2002). But, the continued application of poor quality irrigation water can reduce the yield of farmlands. Water quality for agricultural purpose is determined on the basis of the effect of water on the quality and the yield of the crops, as well as, the effect on the characteristic changes in the soil (FAO, 1985). The most commonly encountered soil problems used as a basis to evaluate water quality are those related to the salinity, water infiltration rate, toxicity and a group of other miscellaneous problems.

Kombolcha is one of the few towns in Ethiopia with a relative greater number of large-scale manufacturing plants including Textile Factory, ELFORA-Meat Processing Factory, Tannery, BGI-Brewery Factory, Steel Product Industry and Flour Factory. On top of this, the town is selected to be an industrial town by Amhara National Regional State of Ethiopia, which indicates the industrial development and its associated pollution risk will increase in the future. The existing industries have been discharging their wastes into the surrounding environment, in particular to the near by river. According to the local woreda agriculture office, more than 25,000 farmers are diverting the effluent contaminated rivers water to irrigate about 2695 ha of farmlands in order to grow different crops including cereals, vegetables and fruits. In addition, the latest report from the local agricultural administration office explains that despite the fact that many farmers and enterprises have used the local rivers for irrigation since long time ago; no study has been conducted yet on the chemistry of the polluted river water for its irrigation suitability (Kalu Woreda Agricultural office, 2007). In Kombolcha, perhaps the most important factor in predicting and managing farmland soil is the quality of irrigation water being used.

The main intention of the study is to provide concrete information on the magnitude of the industrial liquid wastes and help farmers and policy makers to take the necessary corrective measures on time. The impact of industrial liquid wastes on the irrigation water quality was assessed by examining the concentrations of $Na+$, $Ca2+$, $Mg2+$, $BO3-3$, $CO3=$, $HCO3-$, $Cl-$ and values of pH and SAR in the polluted irrigation rivers water through laboratory analysis. Soil samples were also taken to assess the quality of the irrigation water effect on the irrigated farm soils properties.

Material and Methods

Location of Study Area

The study area is found in the town of Kombolcha which is located on the north central part of Ethiopia placed immediately south east of Dessie in the Amahara region at 11°06' north latitude and 39°45' east longitude. River Borkena crosses the town emerging from the east and running to the west direction. In its way all through the town, it receives effluents indirectly through its tributaries rivers named Worka and Leyole. Most of the factories are found closely together in the middle of the town near by the tributary rivers of Borkena.

Methods Samples of irrigation water and farmland soils were collected in three phases with in the irrigation period of the study area. Acceptable standard methods and instrumentations were followed during sample collection procedures.

Sampling Site Selection

Based on the outlining of the irrigation sites and waste disposal points, three areas were selected to take water and soil samples viz a farmland at the above the effluent points (control) which was irrigated by effluent free (freshwater of River Borkena) water and two farmlands below the effluent points which were irrigated by effluent contaminated rivers water (River Leyole and River Worka). The mean values of the parameters in the control fresh irrigation water source and the effluent contaminated water of the other water sources were compared with the widely accepted standards set by FAO.

The soil samples of the respective irrigated farmlands were also considered to assess the extent of the impacts of the effluent contaminated irrigation water on the characteristics the soils. The chemical parameters that have been measured in the diverted irrigation water were also determined from the soil samples of the selected irrigated farmlands. Both surface and subsurface soil samples were taken once from the fresh water irrigated farmland (control or background) at the upper and three times from the effluent mixed irrigated farmlands at the lower of the effluent points through out the irrigation period of the study area. TDS, ESP and SAR were computed following the formulas stated in FAO soil bulletin 42. Chlorides, nitrogen-nitrate, sulfate, chromium and some samples of phosphate were found to be below the detection limit in the first phase samples analyses.

Water Samples

9 water samples were taken from January 2007 to June 2007. The sampling frequency was in three phases throughout the irrigation season. In first phase, additional parameters were analysed other than the mentioned ones in the internationally accepted irrigation water quality guidelines by FAO (1976c; 1985) in order to have a better understanding on the water quality characteristics. The chemical variables analysed in the second and third phase were made to stick only to those recommended in the FAO standard guidelines.

Samples were taken from two sites. One is from a main diversion channel (from the fresh water of River Borkena) above the effluent points. These samples served as background and were non effluent contaminated; the others were from the main channels at lower of the effluent points which were diverted from industrial effluent recipient rivers named Leyole and Worka.

The samples were collected at the same location in all phases of samples collection from the irrigation surface water sources of the selected farmlands. All samples were collected by grab method. While taking the samples, the time of the effluent discharge from the factories was watched out and made the sample collection so in order to take advantage of the effluent presence in the collected irrigation water samples. One liter of water sample was collected per location in a plastic bottle thoroughly cleaned by distilled water. The plastic bottle was rinsed with the water to be sampled just before sample collection and was labelled and recorded on the Water Information Sheet. The samples were then stored in refrigerator at less than 40C temperature till it was delivered to the laboratory for analyses. All samples were transported in ice box and delivered with in two days.

Soil Samples

Before sample collection, site characterization and soil profile descriptions were done by close observation and examination of dug pits on the study areas. First, the surface characteristics were recorded. Then, the soil description was made according to the Guidelines of FAO (1990).

Nine composite surface soils samples and 21 subsurface soils samples were taken from the farmlands irrigated by the above three sources of irrigation water in the irrigation period of the study area. The control soil samples were taken from a farmland placed upper of the effluent points which were irrigated by fresh water (effluent free). The other two areas are found below the effluent point and were irrigated by the effluent contaminated river water of Leyole and Worka rivers.

Composite Surface Soils

Samples were taken from the centre of shovel slice in a 30cm by 30cm core. This was repeated randomly at 20 different spots with in the demarcated farmlands. The collected samples were put in a plastic bucket and thoroughly mixed and at the end, 500gm of soil is removed as the composite sample representing the whole field. The samples were made to air dry for a few days and transported for laboratory analyses in plastic bags. While sampling, areas of back furrows or dead furrows, old fences rows, areas used for manuring or hay storage and livestock feeding, small gullies, slight depressions, terraces, waterways or unusual areas were all avoided.

Subsurface Soil Samples

A pit, in all of the three demarcated sampling areas, was dug to take subsurface soil samples; a depth of 90 cm pits was dug at the selected farmlands. A sample of 500gm soil was removed in each 30cm sections downward. The morphological and other characteristic of the soil was examined in the dug pits which were large enough to allow observations. Sampling from the boundaries of the horizons was avoided. The rule of soil description was made to follow the guidelines of FAO (1996) for soil description. The samples were air dried and transported along with the surface soil samples with in a few days in plastic bags. All soil, surface and subsurface, samples in the plastic bags were labelled and recorded by codes on the Soil Information Sheet.

Physico-chemical Determination of Soil and Water Samples

pH values were read on ORION model SA720 pH meter with a standard solution calibrated at pH values of 4.7 and 9.2. Electrical Conductivity was read on EC meter InoLab (WTW series) which was calibrated using 0.01NKCl standard solution. The cations Na+, K+, Mg+2, and Ca+2 were determined by atomic absorption spectrometer (Varian SP-20). CO3= and HCO3-were measured by titration using phenolphthalein and methyl orange indicators respectively. Chloride was titrated by Argentometry methods. The instrument used for phosphate, nitrate and sulfate measurement was UV visible spectrophotometer. All analyses followed the standard procedures as outlined by USSL staff (1954). TDS and SAR were calculated by formulas as it is suggested in FAO Soil Bulletin 42 (1985).

Data Analysis and Interpretation Techniques

To make irrigation suitability evaluation and quality difference comparison, the values of the chemical variables of lower farmlands irrigation water and soil samples were taken after computing the average of the three phase samples collected in one irrigation period (start, middle and last irrigation times). At the upper farmland (background), a single variables measurement of soil was taken at the starting period of the irrigation season.

These values were used for testing of significant irrigation water quality changes due to industrials effluent discharge in to the irrigating rivers. The most widely applicable irrigation water quality guideline, which is set by FAO, was selected for suitability evaluation. The assumptions made by the selected guideline were then evaluated against the local conditions and it was generally found that most of the assumptions of the chosen guideline for evaluation of irrigation water quality of the rivers are the same to the actual conditions of the study area. There are no as such wide deviations between the assumptions of the guideline and the related local conditions study area. Finally, the values were compared to their respective standards recommended by the internationally accepted guideline in order to evaluate their degree of restriction on use for irrigation.

Since water samples were taken from three different rivers located above and below the effluent points, the test statistics for the significant quality difference in water samples was run by The Independent-Samples T Test.

The absence of irrigation practices at the upstream parts of the wastes draining rivers (River Leyole and Worka) forbids the easiest and rather straight forward quality changes between upstream and downstream water samples due to the intrusion of effluents. Besides, as all the three rivers originate from the same neighbouring catchments areas with more or less the same geological and biophysical characteristics, the quality of the rivers water is assumed to be the same unless otherwise another external element, like the industrial effluents, is introduced in the rivers. To overcome the mentioned limitation, water and soils samples were taken from another neighbouring site with non effluent contaminated river water (River Borkena) and irrigated farm soils. SPSS VERSION-13 software has been employed to run the test. The T Test procedure produces two test of difference between water samples parameters in the two distinct rivers under investigation. One test assumes the variances of each parameter in the two rivers samples are equal.

The Levene Test Statistics tests this assumption. Based on this test, for a significance probability (Sig.) of greater than 0.1, equal variances in the rivers is assumed. Other wise it is ignored and the second test which assumes unequal variance is taken. The frequency of sample variables measurements were three, and the hypothesis was tested at significant level (alpha) 0.05.

Results and Discussions

Quality and Suitability of the Rivers' Irrigation Water: The water samples that have been analysed to measure the levels of electrical conductivity, sodium, chloride, calcium, magnesium, carbonate, bicarbonate, pH and boron.

In addition, sulfate, phosphate, nitrogen-nitrate, fluoride and chromium were added to asses their levels in the first phase of sample collection. The measured values of the parameters were recorded three times over the six months. Some of the parameters, like nitrate, chromium and sulfate, were found to be below the detection limit of the laboratory instruments. Other important parameters like TDS and ESP were computed by the formulas stated in FAO Soil Bulletin 42 (1985). The adjusted SAR (adj RNa) was recalculated using the newer equation adapted from Suarez (1981).

The T test for the pair of upper control water and Leyole irrigation water shows that the mean of Na^+, Cl^-, HCO_3^-, B^{+3} concentrations and the value of EC are greater at the latter. On the contrary, the concentration of Ca^{+2}, $CO_3^=$ and the value of pH were lesser at the latter. Mg^{2+} was found to be the same in both rivers' irrigation water. Statistically, it was seen that there is a significant difference (at P£ 0.05) in mean pH value and Na^+ concentrations between the two sampling locations. Other water parameters (EC, Cl^-, HCO_3^-, SAR), though they indicated appreciable difference in concentrations, they were found to be significantly not different in concentration when compared in the two irrigation water rivers.

The T test for the pair of control fresh water and Worka river irrigation water also reveals that there is a significant (at P£ 0.05) quality difference between them in Na^+, Mg^{+2} concentrations and SAR. The other mean parameters (Cl^-, $CO_3^=$, HCO_3^-, B^{+3}, and pH) were found not significantly different in the two rivers. Chemical parameters, like electrical conductivity, sodium, chloride, bicarbonates, boron, pH, and SAR were found to be at higher concentration in the effluent mixed irrigation water of River Worka in relative to the background effluent free water. The comparison between the two effluents contaminated rivers by the T Test shows all chemical variables but chloride ions are not different significantly (at P£ 0.05).

Effect of Industrial Effluent on Selected Soil Properties of Irrigated Farmlands: as the suitability of water for irrigation is evaluated based on the criteria indicative of its potential to create hazardous soil conditions to crop growth, the effect of the applied irrigation water was referred specifically in terms of salinity, water infiltration, specific ion toxicity and related miscellaneous problems. *I. Salinity:* The mean electrical conductivity of the control irrigation water was 0.4807 dS/m and is put as none restricting for irrigation. The electrical conductivity of Leyole and Worka rivers irrigation water increased to 1.624 dS/m and 1.260 dS/m respectively. Based on the standards of FAO (1985), these figures plunge nearer to a potentially slight degree of restriction to use for irrigation. The salinity of all irrigated farm soil at the upper areas was found to be less than even 0.05 dS/m; this is justified by the low salinity of the applied irrigation water and the practice of surface irrigation methods which help to leach down salts in the rooting depth. However, there was an increase in salinity at the lower area farm soil with 0.0413 dS/m and 0.038dS/m for each effluent contaminated irrigated farmlands as compared to 0.017dS/m of upper fresh water irrigated farmlands. This may indicates that the irrigation water of Leyole and Worka rivers is elevating the salinity of the lower areas farmlands soils.

Water Infiltration: since EC values of all rivers are was not found enough cause permeability problem, the salinity of all irrigation sources is not a factor to cause infiltration problems. The concentration of sodium as compared to calcium and magnesium, which is measured in terms of sodium adsorption ratio (SAR), was found to be less or none restricting in the control irrigation water; however, in the effluent contaminated water of Leyole and Worka rivers, which was detected be 7.71 and 8.18 respectively, it was found high and is potentially restricting. The effect of high SAR water irrigation is noticeable in the soil samples of the irrigated farmlands causing excessive exchangeable sodium percentage (29.24%) in the lower farmlands soils relative to the upper area farm field which has only a maximum ESP of 8.83%.

Specification (Sodium and Chloride) Toxicity: The ions of primary concern were chloride, sodium and boron ion toxicity because these ions are usually related to water toxicity and industrial wastes in arid and semi-arid areas (FAO Soil Bulletin, 1985). But the toxicity effects need to be explained by taking into account indicator crops, which is not the intention of this particular study. However, the assessment of these ions in the water and soils of the irrigated farms could show the general trends with the associated risks of toxicity.

The mean concentration of chloride was quite low in all irrigation water sources and the restriction on use for irrigation is none. The soil samples possessed the smallest (below detection limits) content of chlorides too. At the lower of the effluent points, a little higher reading was obtained in both surface soil samples of Leyole river irrigated farmlands and sub surface soil sample of Worka river irrigated farm fields.

The mean sodium ion concentration of the upper control water of the Borkena river was determined to be less (30.33 ppm/1.32me/

l) and none restriction on use. But the levels in effluent mixed water of Leyole (186.67 ppm/8.11me/l) and Worka (195 ppm/8.48me/l) rivers were higher and can pose moderate restriction on use for irrigation. Accordingly, the soil samples of the downstream farms displayed 160% to 400% increment of sodium ions as compared to the upper farmlands soil samples.

The increased level of sodium at the lower of the effluent point's irrigation water and farmland soils can be attributed to the presence of caustic soda, for the purpose of washing, in the effluents of Kombolcha Textiles, ELFORA Meat Processing and BGI brewery factories. Besides this, the existence of high bicarbonates in the effluent mixed of the two rivers cause Ca+2 and Mg+2 to form insoluble minerals leaving Na+ as the dominant in solution.

At the upper fresh irrigation water of the Borkena river, the mean boron concentration was 0.3 ppm and is none restricting to irrigation. At the lower areas, it was highest (1.15 ppm) in Leyole river which is slight to moderate restriction on use. In the irrigation water of Worka river, it was 0.97 ppm and is near to slightly restricting on use for irrigation (Ayers and Westcot, FAO 1985). All soil samples at the lower areas were below the detection limits, but on areas lower of the effluent points, some samples were indicating that boron is introduced in the surface and subsurface soil of the fields. This could be because of the presence of boric acid in the effluents from the tannery factory.

Miscellaneous: These include measurements of bicarbonate, carbonates, calcium, magnesium and pH of the water. The mean concentration of bicarbonates in Leyole river irrigation water (734 ppm/12.03me/l) was exceptionally high and is beyond the accepted level. But the concentration in control water of the Borkena and Worka rivers irrigation water was 290.67 ppm (4.76me/l) and 367 ppm (6.01me/l) respectively and is in the normal range of concentration for use in irrigation. The bicarbonates ion conditions in the water of the irrigation rivers led to have the same related distribution of bicarbonate content in the soil of the irrigated farm fields with highest accumulation in Leyole river irrigated farmlands (4873.67ppm/79.9me/l) and lowest in farmlands irrigated by the fresh water of the control river. The increased levels of this ion in the soil can be attributed to the long term application of the effluents. The level of carbonates, calcium and magnesium in the irrigation water samples of the three sources.

Leaching Model (Soil)

A leaching model is a hydrological model by which the leaching with irrigation water of dissolved substances, notably salt, in the soil is described depending on the hydrological regime and the soil's properties.

The model may describe the process (1) in time and (2) as a function of amount of water applied.

Leaching is often done to *reclaim* saline soil or to *conserve* a favourable salt content of the soil of irrigated land as all irrigation water contains salts.

Leaching Curves

It shows the soil salinity in terms of electrical conductivity (EC) of the soil solution with respect its initial value (ECi) as a function of amount of water percolating through the soil. The topsoil leaches quickly. The salinity of the deeper soil first increases due to the salts leached from the topsoil, but later it also decreases.

Leaching Efficiency

Owing to irregular distribution of salt in the soil or to irregularity of the soil structure, the *leaching efficiency* (E_L) can be different from unity.

Soils with a low leaching efficiency are difficult to reclaim. In the Tagus delta, Portugal, the leaching efficiency of the dense clay soil was found as low as 0.10 to 0.15. The soil could not be developed for intensive agriculture and was used for rearing of bulls in coarse natural pasture.

The clay soil in the Nile delta, Egypt, on the other hand has a much better leaching efficiency of 0.7 to 0.8. The observed values of soil salinity correspond best to a leaching efficiency of about 0.75. The figure illustrates the calibration process of leaching efficiency, which parameter is difficult to measure directly.

Leaching Requirement

The leaching requirement may refer to:

- The total amount of water required to bring the soil salinity from an initially high value down to an acceptable value in accordance with the salt tolerance of the crops to be grown. From figure 1 it is seen that 800 mm of water (or 8000 m^3/ ha) is required to bring the soil salinity down to 60% of its original value in the soil layer at 40 to 60 cm depth. When the salinity must be less than 60%, extrapolation of the leaching curve, the use of a *leaching equation* or a leaching model like Salt Mod is necessary to obtain a reliable estimate of the additional leaching requirement.

- The annual amount of percolation water (i.e. the extra amount of irrigation water on top of the crop consumptive use) required to conserve an acceptable salt balance of the soil in accordance with the salt tolerance of the crops to be grown. The ratio

F_L = Perc/Irr, where Perc = amount of required percolation water, and Irr = total amount of irrigation water, is called *leaching fraction*.

Leaching Equation

The downward limb of the leaching curves, as in figure 3, can be described with the leaching equation :

- $Ct = Ci + (Co-Ci) \exp (-E_L.T.Qf/Ws)$

where C = salt concentration, Ct = C in the soil at time T, Co = C in the soil at time T=0, Ci = C of the irrigation water, E_L = leaching efficiency, Qp = average percolation rate through the soil, and Ws = water stored in the soil at field saturation.

Leaching Fraction

To conserve an acceptable salt balance of the soil in accordance with the salt tolerance of the crops to be grown, the leaching fraction must be at least:

- $F_L = Ci/Cs$

where Ci = salt concentration of the irrigation water, and Cs is the acceptable salt concentration of the soil moisture at field capacity in accordance with the salt tolerance of the crops to be grown.

Plants Thrive on Varying Levels of Soil Moisture

Every plant has a preference when it comes soil moisture. Tags will tell you the plant prefers a "moist" soil, or a "well-drained" soil, some even a "dry" soil, but how do you know what moist is and how is moisture measured?

Soil moisture is the amount of water present in the soil surrounding the roots on a continual basis; not right after a heavy rain or a spring snow melt, but rather on a day-to-day, month-to-month average. It varies greatly due to soil type, amount of organic matter present, exposure to sunlight and temperature.

But what doesn't vary is the plant's dependence on this moisture; not only do the roots take up water, but important minerals which have become dissolved in the water. We know water is critical for photosynthesis, the plant's food-making process. We also know moisture affects seed germination. Some seeds require moisture to help break down their seed coat so germination can occur; morning glory seeds must be soaked in water for 24 hours prior to planting. Others require moisture for survival immediately following germination, such as the spores of a fern. Think about the environment you will find ferns thriving in; it's continually moist.

How does your soil rate when it comes to holding moisture? Soil labs often determine the moisture capacity of a sample by weighing it, heating it until the moisture has evaporated and then re-weighing; the difference is calculated as the percent of soil moisture. For homeowners, a simpler technique known as the "big squeeze" works well. Dig into the soil about 6 inches, grab a handful and squeeze.

If it remains in a "ball" and a wet outline of water appears in your palm, your soil is considered very moist. Should the "ball" break apart but remain in large clumps, your soil is moist.

A well-drained soil will "ball" and then crumble apart. A dry soil, depending on the soil type, will either flow through your fingers like sand through an hourglass or be rock hard and difficult to chip out of the ground. As mentioned above, many other factors contribute to a soil's moisture-holding capacity, but one thing we as gardeners can do to help our soil is to add organic matter and lots of it.

Sandy soil is made up of larger soil particles than a clay soil. As a result, when rain falls it leaches or drains through the sand very quickly, thus sand is noted as having a poor moisture-holding capacity.

Clay, with its very tiny soil particles, is able to hold the initial rainfall for a longer period of time than sand, but when it dries out it is rock-hard. Organic matter also known as decomposing plant and animal debris such as straw, grass clippings, peat moss, leaves, compost and manure are able to hold 20 times their weight in water.

By adding organic matter to our soils, we not only add nutrients but help the soil retain moisture. Over time, if we add organic matter on an annual basis, we will begin to notice the quick drain of sandy soil beginning to slow and the fine particles of clay soils beginning to mix with the larger pieces of decaying matter. Soils high in organic matter are known as loam.

When it comes to hanging baskets, professional growers use a potting mix formulated for moisture retention. This is an example we should follow when filling our window boxes and planters. But because these planters and baskets are exposed to the sun's drying rays, keeping them moist can be a chore.

During the past few years, a product has come on the market that can be added to the mix to help retain moisture. It's similar to the filling you find in disposable diapers and can be found under various trade names at local home and garden centers, one being Soil Moist.

As little as a half-teaspoon of these tiny round beads, depending on the container size are added to the potting mix. When watered, they swell and turn into little gelatin globs that can hold excess moisture until the soil's moisture level drops, then they release the water they're holding.

Here are a few other plant survival tips for the summer heat:

- Wilted plants in containers should be immersed in a bucket of water until air bubbles stop rising. Often the pots become so root-bound or the potting mix too dry to absorb water from above.

- If leaving for a long weekend, place planters in a shady location and water thoroughly just before leaving. For longer jaunts, ask a friend to tend to them in your absence.

- Finally, a 3-inch layer of mulch works wonders in helping to retain soil moisture by interfering with evaporation and controlling soil temperature fluctuations.

4

Irrigation Impact on Agricultural Growth and Poverty Alleviation

Though the positive impact of irrigation on agricultural intensification and increased crop yield has been very well documented by various studies, the marginal returns of irrigation versus other factor inputs such as, farm technology and other rural infrastructure development are still a controversial issue in rural development literature.

Improved information and understanding of the scale of incremental benefit of irrigation and other factor-inputs to agricultural growth and development and to poverty alleviation process has large public policy implications on setting rural development policy in a region. This has particularly more relevant in setting irrigation and agricultural investment and financing policies. This information is also important considering the recently increased global public policy priorities and thrusts on poverty reduction strategies.

In this study an attempt is made to analyse the incremental impact (input specific effects) of irrigation and other factor inputs on growth of agricultural productivity of all inputs taken together (i.e., Total Factor Productivity) and their implications on poverty alleviation in India over the last two and half decades. The Total Factor Productivity is also called as productivity of all inputs taken together, and it is different than the conventionally understood productivity measures like crops yield, water productivity or labour productivity. In addition, the study also examines the structure and relative importance of the factors that affect the over time variation in poverty and rural consumption level across the states in India. This is done by using annual time series and cross section (state) data analysis technique across 14 major states of India covering the period from 1970 to 1994, which covers more than 90 percent of the agrarian economy of India.

The overall growth and technical change in the agricultural sector has large implications on expanding the economic base and poverty alleviation process in a region. Past empirical studies have shown that ultimately the growth in productivity of all factors (TFP) in agriculture is the backbone for alleviating rural poverty in developing countries. While summarizing the previous literature on agriculture growth and poverty reduction, Mellor (2001) points out that agricultural growth has a profound impact on poverty reduction in the developing countries including the reduction of the inequity over time.

The actual level of impact of agricultural growth on poverty in fact varies by the nature, region and time period selected for the studies. Though most of the previous studies have unequivocally demonstrated that agricultural productivity growth has a profound impact on reducing poverty in Asia, the existing literature on rural poverty has failed to examine the incremental impact of each of the factor inputs on agricultural productivity growth as well as their marginal impact on poverty alleviation, and rural income enhancement.

Several irrigation impact related case studies and regional studies in India have illustrated that irrigation management has a profound role to play in the poverty alleviation process. Some of the recent aggregate level empirical studies in India have also shown that irrigation access has a positive impact on poverty reduction. However, there is no straightforward relationship shown between irrigation and poverty alleviation and the irrigation impact on poverty alleviation depends upon several other intermediate factors. Thus an improved understanding of the structure of the impact of various factors and a quantification of the marginal impacts of each of the factor input on poverty measures is important for developing efficient and effective policy instruments in poverty alleviation and rural development programs in general..

Results and Discussions

Factors Affecting Productivity of all Inputs

This study quantifies the marginal impact of irrigation and other factor inputs on the agricultural productivity of all inputs and on the two key poverty measures across the states. The empirical results show that there is no significant growth taking place in the agriculture productivity level when the level of all inputs use and their costs are taken together (in terms of economic and technical efficiency) in India over the last two decades. This productivity growth of all inputs is different than the simple crops yield or labour productivity.

This means that one percent increase in irrigated area has increased about 0.32 percent in the productivity of all inputs (TFP) in India during 1970-1994. This is a very high impact on agricultural productivity when compared to the impact of other factors such as fertilisers, HYV and road infrastructure, where the elasticity varies only from 0.04 to 0.09. The impact of rural literacy rate (percent of rural population) is positive and statistically highly significant in explaining the over time variation in agricultural productivity of all inputs (TFP). The marginal impact of rural literacy on agricultural productivity is the largest among the variables selected for the analysis. This large impact of rural education is possible considering the fact that technical change and agricultural productivity (represented by increased TFP index) and rural development are directly related to the adoption of improved technology, selection of appropriate mix of crops and inputs, and timely application of these inputs, including the farmers' ability to effectively process market and price information and farm managerial decisions. The impact of variable road infrastructure is also positive and significant which may be capturing the effects of market access in agricultural and rural development.

Factors Affecting Poverty Rate and Per Capita Rural Consumption Level

The study has also analysed the direct impact of selected factor inputs in explaining the variation in poverty measures (poverty measure in head count ratio and rural per capita consumption) across the states for 1970-1993, using the same set of factors used on analysing the agricultural productivity. The study found that the extent of rural poverty was unequivocally higher in a state with less extent of irrigation especially in the early 1970s.

Moreover, the relationship between rural poverty and irrigation has been decreasing in the recent past. This shows that the scale of poverty level was very severe in the early 1970s, with more than 60 percent of the rural population under poverty line (head-count ratio) in almost all parts of central and eastern India. Irrigation development was also very low in these regions at that time. However, the situation has improved in the early 1990s where rural poverty is mostly concentrated in states like Bihar, Orissa, and Madhya Pradesh, where irrigation development is poor as of today. While analysing the independent relationship between irrigation and rural poverty, there appears to be a strong inverse relationship between the incidence of rural poverty and percentage of gross cropped area irrigated.

Besides analysing the role of irrigation in rural poverty through regression analysis, we have also depicted the relationship between irrigation and poverty in the two graphs. The trend on variation in irrigation and the various measures of poverty, and how they have changed overtime is illustrated. The level of irrigation has increased more than double between 1960 and 1990. All the poverty measures have declined unequivocally during these periods. The Head Count Index (HCI) which measures the percentage of population below poverty line using consumption expenditures has reduced over the period of the last three decades. More importantly, the other two measures of poverty, poverty gap index (PGI) and Foster-Greer-Thorbecke (FGT) have declined at faster rate over the period of last three decades. This means all three measures of rural poverty in India have declined during the period, largely contributed by the success of irrigated agriculture over the years. As depicted through figures, the regression results also clearly demonstrate the role of irrigation in reducing the rural poverty.

The negative sign of time trend variable in poverty model, which shows overtime change on trend of poverty rate, suggests that poverty level in India has unequivocally decreased during the time period of 1970-1993. This is also supported by the positive sign of this time trend variable in the consumption model which shows over time increasing rate of per capita consumption of rural population. Among all the variables selected for analysing the poverty measures in this study, the irrigation factor has the strongest influence in explaining the reduction in poverty.

Irrigation has even a larger marginal impact on reducing the poverty than the impact of rural literacy. Whereas rural education factor was earlier found to be strongest in influencing the agricultural productivity across the states. Likewise, the increased HYV adoption and fertilizers use have also played a favourable role in reducing poverty in India, but their influence on poverty reduction is lower than the marginal incremental impact of irrigation and rural literacy. Unlike in productivity growth, the variable road infrastructure does not play any positive and favourable role in explaining the variation in rural poverty in India during the time period selected for this study.

India's Groundwater Challenge

Groundwater is a dry topic unless you happen to have a dry well. That said, groundwater management involves some of the most complex and socially challenging sets of issues facing India in the 21st century. Furthermore, how those issues are resolved will affect both the environment and the day to day life of most people living in rural and urban areas.

Groundwater is an invisible resource. As a result, both the dynamics of the resource base and the services it produces are often poorly understood. This article focuses first on the broad array of environmental and social services that depend on groundwater. Key aspects of the resource base and emerging problems are presented next. The final section outlines some of the social, ethical and institutional challenges inherent in sustainable management.

Over the past 50 years, expansion of groundwater irrigation has played a lead role in food security. Yields in areas irrigated by groundwater are often substantially higher than yields in areas irrigated from surface sources. In India, for example, research indicates that yields in groundwater irrigated areas are higher by one third to one half than in areas irrigated from surface sources and as much as 70-80% of India's agricultural output may be groundwater dependent.

Higher yields from groundwater irrigated areas are due, in large part, to its ease of control and reliability. Early studies indicated that water control alone can reduce the gap between potential and actual yields by about 20%. This translates into substantial benefits. Reliability is even more important. Groundwater is a key buffer against drought and normal variations in rainfall. Overall, increased yields from groundwater irrigated areas have translated into substantially higher yields and are thus a major factor in food production at the regional and national levels.

Furthermore, some of the most important food security benefits related to groundwater lie at the level of individual farmers. The vulnerability to natural hazards of different groups in society, including those that threaten food security, can be explained by their access to networks of key productive and social resources. For rural populations, groundwater is among the most important of these resources.

Households with access to key resources are able to build support systems that reduce their vulnerability to natural hazards. Groundwater irrigation reduces the risk that investment in labour, seed, fertilisers, pesticides and other inputs will be lost due to drought or the variability of precipitation in normal years while higher yields enable households to generate surpluses. As a result, households with access to groundwater tend to have higher levels of savings and are able to make investments in other productive resources or activities.

When drought strikes or there is a gap in rainfall these households have a dual advantage. First, they are far less likely to suffer losses than those without access to groundwater. Second, even if 'the well runs dry', households that own wells have often been able to save cash or food and invest in alternative sources of income. As a result, they have assets that can carry them through periods of scarcity or crisis.

Groundwater is a highly important source of domestic water supply. In India, roughly 80% of rural water supply for domestic uses is met from groundwater. The importance of potable drinking water is clear. As in the case of other uses, however, this is only a portion of the value of groundwater as a source of domestic supply. Wells in villages and towns free people, particularly women, from long daily walks to fetch water from springs or rivers for livestock and domestic uses. This frees time and labour for other activities. Furthermore, since water no longer has to be carried over long distances, more is often used. This can have major health benefits. In addition, because of the filtering nature of the soil and frequent long residence time underground, groundwater is commonly much cleaner than surface sources.

Groundwater is a key resource for poverty alleviation and economic development. Evidence indicates that improved water sources generate many positive externalities in the overall household micro-economy. In areas dependent on irrigated agriculture, the reliability of groundwater sources and the high crop yields generally achieved as a result often enable farmers with small landholdings to increase income. In India small and marginal farmers (those having less than 2 hectares) own 29% of the agricultural area. Their share in net area irrigated by wells is, however, 38.1 % and they account for 35.3 % of the tubewells fitted with electric pump sets.

Thus, in relation to operational area, small and marginal farmers tend to have proportionally more irrigated land than larger farmers. With productivity on irrigated lands being much higher than that on non-irrigated tracts, better access to irrigation for small and marginal farmers can significantly reduce poverty.

The positive economic impact of groundwater development extends beyond well owners. Access to groundwater stabilises the demand for associated inputs, and leads to the spread of support services for pumps and wells, creating a base for small scale rural industries. Furthermore, the spread of groundwater irrigation can increase demand for labour. In India, for example, labour accounts for approximately 44 % of the cost of installing a well and the additional indirect employment created on every hectare of irrigated land through increased agricultural activity is approximately 45 days per hectare. Therefore, the expansion of groundwater irrigation has significant ripple effects, creating employment throughout rural economies.

The equity impacts of groundwater development for irrigation are, however, not all positive. Modern tubewell and drilling technology tends to be capital intensive. As a result, early exploiters of groundwater have typically been large farmers who produce surpluses for the market Small holders growing subsistence crops often depend on supplementary groundwater irrigation using a variety of man and animal-driven water lifting devices from shallow, open wells. The expansion of energised pumping technologies tends to draw water levels down, driving shallow wells and muscle-driven devices out of business. This was, for instance, the case throughout the Gangetic basin and other parts of India during the 1960s.

As water levels began to decline, some state governments in India attempted to implement administrative regulations, such as selective credit controls, restrictions on electricity connections and siting and licensing rules. These regulations did not affect landowners who were able to install tubewells early, but limited the entry of latecomers – particularly the resource poor who depended on credit or access to subsidised electricity in order to afford the capital cost of operating and installing pumps. In addition, the economically and politically powerful were generally able to bypass regulations via 'adjustments' with officials or by depending on their own financial resources for well construction and operation.

Equity considerations are generally a major point of tension as management needs emerge. Rapid unrestricted development of groundwater has reduced poverty by giving the poor access to a key resource for production. This same pattern of unrestricted development, however, is the primary cause of over-extraction and quality problems now emerging in many parts of the world. As groundwater problems grow, marginal populations are often the first affected. Water level declines, for example, have the largest economic impact on individuals who are unable to afford deeper wells – i.e. the poor. The poor are also the least well positioned to protect their interests if groundwater extraction must be reduced. Restrictions on new wells tend to affect them much more than wealthy communities where wells were installed much earlier. Wealthy individuals and communities are also often able to work around these and other types of management restrictions while poorer communities (who generally lack political as well as economic leverage) have less ability to do so. In sum, there is an inherent tension between equitable access to groundwater for all sections of society and sustainable management of the resource base.

The array of environmental services or values dependent on groundwater are often poorly understood. Environmental concerns related to groundwater generally focus on the impacts of pollution and quality degradation on human uses, particularly domestic supply. Development impacts on the groundwater environment are, however, different from the numerous environmental services provided by groundwater resources in their natural state.

What are some of the most important environmental services provided by groundwater? The following list, while not exhaustive, illustrates the degree to which many environmental values are dependent on groundwater:

- *Catchment base flow* derived from groundwater discharge is perhaps the most evident environmental value associated with groundwater. In many areas springs and the dry season flow in rivers depend heavily on groundwater.

- *Instream fisheries and aquatic ecosystems*: Instream flows are critical for the maintenance of fisheries and aquatic ecosystems. As previously noted, groundwater contributions can be a dominant source of water for instream flows, particularly during droughts and dry seasons.

- *Inland wetlands*: Wetlands are some of the most productive and biologically diverse inland ecosystems. In many, if not most, cases water availability in wetlands depends on high groundwater levels.

- *Surface vegetation*: Groundwater levels directly influence many vegetation communities. Phreatophytes, plants that derive a major portion of their water needs from saturated soils, can be the dominant vegetative species in ecosystems where groundwater levels are shallow. They often form critical wildlife habitat and may serve as important sources of food, fuel and timber.

In many areas the environmental values dependent on groundwater conditions are closely intertwined with a broad array of human use patterns. Groundwater is an integral part of linked hydrologic, ecologic and human use systems. Changes in surface water use, groundwater use or vegetation can send ripple effects throughout these inter linked systems – often with effects that are difficult to predict.

Throughout much of South Asia, groundwater development has proceeded at an exponential pace over the past four decades. In India, the number of shallow tubewells doubled roughly every 3.7 years between 1951 and 1991. Rapid development has engendered its own set of issues.

In many arid and hard rock zones, overdraft and associated quality problems are increasingly emerging. Although the area currently affected by groundwater overdraft may be limited, blocks classified as dark or critical increased at a continuous rate of 5.5% over the period 1984-85 to 1992-93. If continued at this rate, the number would double every 12.5 years. This implies that by the year 2017-18 (25 years from 1992-93), roughly 1532 blocks or 36% of the 4248 blocks in the listed states would be dark or critical.

Overdraft is, however, only a fraction of the management challenge associated with groundwater. Large areas, particularly in the command of surface irrigation systems suffer from waterlogging and associated salinity or alkalinity problems. Furthermore, development impacts on the environment and non-agricultural users can be major even where overdraft or waterlogging are absent. Seasonal watertable fluctuations can affect shallow wells, low season flows in surface streams, and pollution loads. The impact of this on drinking water availability, the poor and the environment can be major.

Furthermore, it is important to recognize that *overdraft and water level declines typically affect the sustainability of uses that are dependent on groundwater long before the resource base itself is threatened with physical exhaustion.* Many uses and environmental values *depend on the depth to water* – not the volume theoretically available. In the case of the Ganges basin, for example, water level declines would exclude the poor from access to groundwater (due to the cost of increasing well depth) and would reduce base flows in streams long before the aquifer would face any threat of depletion. The Ganges basin contains, in some locations, over 20 thousand feet of saturated sediment. Dewatering of only the top few tens of feet would, however, have tremendous economic and environmental impacts.

Beyond overdraft and water level declines lie the questions of water quality and pollution. Pollution or quality declines can cause reductions in water availability that are far less reversible than overdraft. Non-point source pollution from agriculture and other sources combined with point source pollution represents a major management challenge. Furthermore, not all quality problems are human induced. Probably the most extensive case of arsenic poisoning from groundwater is that of Bangladesh and West Bengal. Arsenic occurs naturally in the ground waters abstracted from the alluvial deltaic sediments of the Ganges-Brahmaputra-Meghna river systems and an area around 75,000 km2 is thought to be affected by groundwater with high arsenic concentrations.

Despite the widespread nature of emerging problems, how extensive groundwater mining and pollution problems really are remains uncertain. Official statistics on the number of blocks where extraction is approaching or exceeds recharge may be misleading since great uncertainty exists over the reliability of published extraction and recharge estimates. Furthermore, even the basic water table measurements on which these estimates rest may, in some cases, be open to question. Uncertainty is also inherent in relation to the extent of pollution. Clearly pollution loads have increased substantially over recent decades with increases in the use of agricultural chemicals, industrial discharges and urban waste. At the same time, no comprehensive data sets are available that would allow identification of the extent and distribution of different pollutants. Quality problems are also often identified 'after the fact'. Again the arsenic case is illustrative. This was only recognized as a widespread problem when large scale cases of arsenic poisoning began to appear.

A critical challenge in interpreting both quantity and quality problems relates to understanding of the resource base and its dynamics. Many individuals, groundwater professionals included, conceptualize groundwater as flowing smoothly through the earth with rapid recharge from rainfall and relatively uniform water quality. In reality, however, complex rock formations and differential recharge rates result in far more complicated dynamics. These, in turn, greatly complicate understanding of resource conditions. Two examples are illustrative. In Rajasthan, local communities often suggest that groundwater overdraft problems are alleviated whenever rainfall is above normal. In many cases, however, the water they pump has been underground for hundreds and, in some cases, as much as 20,000 years. Recharge dynamics depend on the permeability of soils and flow patterns are often only weakly related to short term fluctuations in precipitation. Similarly, major quality problems can depend heavily on very localized conditions. Arsenic, for example, is preferentially mobilized under reducing conditions. As a result, the amount of arsenic encountered in water from a given well can depend on the amount of organic matter available precisely where the well was drilled. A well that happens to pass near a tree trunk buried deep underground may have substantially more arsenic than a well drilled only a short distance away.

Debates over the need for management are often clouded by uncertainty over the extent and nature of emerging problems. At the same time, delays in initiating management can have irreversible implications. Once polluted, groundwater aquifers are often impossible to remediate. Pollutants can become attached to the aquifer matrix (the sand, soil and rocks) and serve as a continuous source of contamination. Overdraft can lead to aquifer compaction (reduction of pore spaces) making recharge impossible. Even if recharge is technically feasible and sufficient water is available, the low flow rates characteristic of many fine grained aquifers can make the process so slow that little can be done on a human time scale.

As a result, overdraft is often equivalent to mining a resource such as oil that can not be replenished. Even experts often have difficulty identifying the emergence of irreversible problems until after the fact. Managing uncertainty is thus a major part of any groundwater management equation. Groundwater is an invisible, poorly understood resource, yet one that is critical to a wide variety of social, economic and environmental services. Pollution and declining water levels represent direct threats to the sustainability of environmental, domestic, agricultural and industrial uses dependent on groundwater. In addition, as demands grow and the limits of sustainable extraction become evident, competition between agricultural and other users is increasing rapidly.

This can generate competitive extraction between individuals. Each person extracts as much groundwater as possible in order to capture benefits for themselves before the resource is exhausted. The net result can be a spiral of growing demands and decreasing availability. Competition is, thus, a critical social issue that must be addressed in order to manage groundwater on a sustainable basis.

Technical options for resolving competition over groundwater are generally limited. In most cases communities and groups affected by groundwater overdraft advocate aquifer recharge. Throughout India the real viability of this option is often limited. In much of Gujarat, for example, estimates indicate that increases in recharge could only reduce overdraft by 10% and in many other areas little unutilized surface water is available that could be recharged. Even where water is available, the timing and distribution of rainfall limit the ability to recharge aquifers. Recharge is a slow process limited by the infiltration rates of water through soil and underlying formations. Precipitation, in contrast, is often highly seasonal and occurs in brief bursts little of which can be stored for recharge.

Beyond the technical limitations in recharging aquifers, three concepts are central to understanding the social challenges inherent in resolving competition over groundwater resources. These are:

- *Interdependency*: Groundwater is a lynchpin that links and creates points of interdependency between agricultural, environmental and economic systems. Environmental values, access for the poor to water, food security during drought years and the economic viability of different crops may, for example, all depend on the maintenance of specific water table or water quality conditions. These conditions are, in turn, often dependent on water use patterns. Recharge from 'inefficient' surface irrigation systems, for example, often helps to maintain high groundwater levels and, through that, base flows in rivers, groundwater access for the poor and so on.

- *The Public Good Nature of Many Groundwater Services*: Individual users can only capture the 'extractive' values associated with groundwater, i.e. products produced by pumping and using it in a specific application. This extractive value does not, however, reflect the environmental, drought buffer and other services produced by groundwater when it is left in the aquifer. These services are *public goods* – while individuals may benefit from them, the conditions depend on the cumulative actions of all users. There are strong economic incentives for individuals to overpump groundwater or ignore their contribution to pollution of an aquifer. This leads to chronic undervaluation of groundwater when it is sold in water markets or analysed using standard economic approaches.

- *Scale*: In most cases, groundwater cannot be managed at a very local scale. Aquifers generally extend under regions encompassing anywhere from tens to thousands of villages. As a result, aquifer dynamics limit the impact individuals or villages can have on groundwater conditions. At the same time, approaches to management implemented through state agencies are often difficult to adapt in ways that reflect local or regional variations in groundwater conditions and use.

In some parts of the world, notably the United States, the approach to groundwater management hinge on a combination of private rights and regulatory mechanisms. This type of approach attempts to resolve competition by providing a measure of security for all users by specifying individual use rights and using regulation to address public good aspects. In some cases groundwater rights are established that specify the volume individuals are allowed to pump, the types of uses permitted and whether or not the water can be sold to other users. Water markets are widely advocated as a mechanism for ensuring water is allocated to the highest value uses wherever transferable water rights have been established.

Approaches based on regulation and the establishment of individual water rights and water markets face tremendous challenges in India. On a purely practical level, how groundwater rights could be established, monitored and enforced given the millions of wells and conditions in rural India, is far from clear. Regulation would also be difficult and probably highly inequitable in practice. A model bill for groundwater regulation was initially circulated by the Central Ground Water Board in the early 1970s. It proposes a highly centralized system of regulation by state agencies. Modified versions of this have now been adopted in a few locations in India but little implementation has actually occurred.

Beyond the practical limitations of approaches based on individual rights and regulation by the state, however, it is important to recognize the inherently incomplete nature of approaches based on rights, regulation and water markets. It is difficult to define water rights (whether allocated to individuals or communities) in ways that capture the interdependent and public good nature of the services produced by groundwater. When rights are transferable through market mechanisms, the values reflected in the market still tend to be the direct use values rather than the public goods. When attempts are made to regulate groundwater use and transfers in ways that protect public goods approaches rapidly become complex and inflexible – unable to respond to the diverse conditions encountered at local levels or to the dynamic nature of conditions. Because of the above limitations, approaches to groundwater management need to reflect the political nature of such decision making in order to be effective. The public good nature of groundwater services and inter linked and often poorly understood character of systems dependent on groundwater tends to generate substantial debate over management approaches and objectives. How this competition is resolved generally depends on the ability of different groups to first understand the nature of emerging problems and management options and second to insert their views into the decision making process that determines management actions.

Access to information, economic power and the ability to organize greatly influence the ability of different groups to effectively engage in this type of dialogue. Under current conditions, groundwater management decisions are effectively based on economic power – the ability of individuals to afford the costs of deepening their own wells and keep on pumping. Resolving this probably requires approaches based on balance of power concepts and explicit recognition of the political nature of management needs.

Key points of leverage that may encourage the development of effective groundwater management systems could include:

- improved access to information;
- legal standing for groups and individuals to force protection of public interest values; and
- the creation of management organizations capable of functioning at an intermediate scale (i.e. between the village and the state).

Irrigation in India

Except in southeastern India, which receives most of its rain from the northeast monsoon in October and November, dryland cultivators place their hopes for a harvest on the southwest monsoon, which usually reaches India in early June and by mid-July has extended to the entire country. There are great variations in the average amount of rainfall received by the various regions—from too much for most crops in the eastern Himalayas to never enough in Rajasthan. Season-to-season variations in rainfall are also great. The consequence is bumper harvests in some seasons, crop-searing drought in others. Therefore, the importance of irrigation in India cannot be overemphasized.

Irrigation i India has been a high priority in economic development since 1951; more than 50 percent of all public expenditures on agriculture have been spent on irrigation alone. The land area under irrigation expanded from 22.6 million hectares in FY 1950 to 59 million hectares in FY 1990, an increase of 161 percent in four decades. This increase was about 33 percent of the estimated potential. The overall strategy has been to concentrate public investments in surface systems, such as large dams, long canals, and other large-scale works requiring huge outlays of capital over a period of years, and in deep-well projects that also involve large capital outlays. Shallow-well schemes and small surface-water projects, mainly ponds (called tanks in India), have been supported by government credit but were otherwise installed and operated by private entrepreneurs. Roughly 42 percent of the net irrigated area in FY 1990 was from surface water sources. Tanks, step wells, and tube wells provided another 51 percent; the rest came from other sources.

Between 1951 and 1990, nearly 1,350 large-and medium-sized irrigation works were started, and about 850 were completed. The most ambitious of these projects was the Indira Gandhi Canal, with an anticipated completion date of close to 1999. When completed, the Indira Gandhi Canal will be the world's longest irrigation canal. Beginning at the Hairke Barrage, a few kilometers below the confluence of the Sutlej and Beas rivers in western Punjab, it will run south-southwest for 650 kilometers, terminating deep in Rajasthan near Jaisalmer, close to the border with Pakistan. A dramatic change already had taken place in this hot and inhospitable wasteland by the late 1980s. As a result, desert dwellers switched from raising goats and sheep to raising wheat, and outsiders flocked in to purchase six-hectare plots for the equivalent of US$3,000.

Progress in irrigation has not been without problems. In India, arge dams and long canals are costly and also highly visible indicators of progress; the political pressure to launch such projects was frequently irresistible. But because funds and technical expertise were in short supply, many projects moved forward at a slow pace. The Indira Gandhi Canal project is a leading example. And the central government's transfer of huge amounts of water from Punjab to Haryana and Rajasthan, frequently cited as a source of grievance by Sikhs in Punjab, contributed to the civil unrest in Punjab during the 1980s and early 1990s.

Problems also have arisen as ground water supplies used for irrigation face depletion. Drawing water off from one area to irrigate another often leads to increased salinity in the supply area with resultant effects on crop production there. Some areas receiving water through irrigation are poorly managed or inadequately designed; the result often is too much water and water-logged fields incapable of production. To alleviate this problem, more emphasis is being placed on using irrigation water to spray fields rather than allowing it to flow through ditches. Furthermore, charges of corruption and mismanagement have been levied against government-operated facilities. Cases of bribery, maldistribution of water, and carelessness are frequently raised in the media.

Another major problem has been the displacement of thousands of people, usually poor people, by large hydroelectric projects. Critics also claim that the projects are damaging to the ecology. Smaller projects and such traditional methods for irrigation as tanks and wells are seen as having less serious impact. In the late 1980s and early 1990s, the debate between large-scale versus small-scale projects came to the fore because of the US$3 billion Sardar Sarovar project on the Narmada River. Sardar Sarovar, as conceived, was one of the world's largest hydroelectric and irrigation projects. Some 37,000 hectares of land in Madhya Pradesh, Gujarat, and Maharashtra were slated to be submerged following the construction of some 3,000 dams, 75,000 kilometers of canals, and an electric power generating capacity of 1,450 megawatts of power per year. Included among the 3,000 dams was the proposed 160-meter-high Sardar Sarovar Dam. In 1985 the World Bank agreed to loan US$450 million for the project. Environmentalists in India and abroad, however, argued that the project was ecologically undesirable. In the face of this strong protest, the World Bank appointed a two-member team in 1991 to review the project. Despite a negative review of the environmental impact by the team, World Bank funding and the project continued. By 1993, however, in the face of continued international protest as well as opposition and a call for a satyagraha by villages in the affected areas, the central government cancelled the dam project loan. Work on the Sardar Sarovar project continues, however, with funds provided by the central government and the governments of the three states involved.

Although India had the second largest irrigated area in the world, the area under assured irrigation or with at least minimal drainage is inadequate. The irrigation potential estimated to have been created by the early 1990s was about 82.8 million hectares. This amount includes the gross irrigated area plus the potential for double cropping provided by irrigation. There was a cumulative gap in irrigated land use of about 8.6 million hectares until FY 1990, by which time the gap had decreased through improved land management.

Irrigation Water Requirements

The assessment of the irrigation potential, based on soil and water resources, can only be done by simultaneously assessing the irrigation water requirements (IWR).

Net irrigation water requirement (NIWR) is the quantity of water necessary for crop growth. It is expressed in millimetres per year or in m^3/ha per year (1 mm = 10 m^3/ha). It depends on the cropping pattern and the climate. Information on irrigation efficiency is necessary to be able to transform NIWR into gross irrigation water requirement (GIWR), which is the quantity of water to be applied in reality, taking into account water losses. Multiplying GIWR by the area that is suitable for irrigation gives the total water requirement for that area. In this study water requirements are expressed in $km^3/$ year. Calculations of irrigation water requirements are done while preparing national water master plans or irrigation projects. Useful information was obtained from a number of country studies available from AQUASTAT, but the information was based on many different approaches. For the purpose of this study the need was felt to develop a method of computing irrigation water requirements for the whole continent in a systematic way. In order to be able to do this at the scale of the continent, assumptions have to be made on the definition of areas to be considered homogeneous in terms of rainfall, potential evapotranspiration, cropping pattern, cropping intensity and irrigation efficiency.

Methodology

For the calculation of irrigation water requirements the following steps have been followed:

- Delineation of major irrigation cropping pattern zones. These zones are considered homogeneous in terms of types of irrigated crops grown, crop calendar, cropping intensity and gross irrigation efficiency. Represented on the map of Africa, they should be viewed as regions where some homogeneity can be found in terms of irrigated crops. The cropping pattern proposed for the zone should be viewed as representative of an 'average' rather than a 'typical' irrigation scheme.

- Definition of the area of influence of the climate stations (in GIS) and quality check on the climate data.

- Combination of the irrigation cropping pattern zones with the climate stations' zones (in GIS) to obtain basic mapping units.

- Calculation of net and gross irrigation water requirements for different scenarios.
- Comparison with existing data and final adjustment.

Delineation of Irrigation Cropping Pattern Zones

The criteria used for the delineation of the irrigation cropping pattern zones were, in order of decreasing importance: distribution of irrigated crops, average rainfall trends and patterns, topographic gradients, presence of large river valleys (Nile, Niger, Senegal), presence of extensive wetlands (the Sudd in Sudan), population pressure, technological differences and crop calendar above and below the equator (Zaire). The starting point was the type of irrigated crops currently grown in Africa. This resulted in 18 zones. From these zones, sub-zones showing a different cropping intensity or a different crop calendar were defied. This resulted in a total of 24 irrigation pattern zones, which are considered to be homogeneous for:

- crops currently grown;
- crop calendar;
- cropping intensity.

Only the main crops currently grown, those occupying at least 85% of the irrigated area, were considered. Land occupation of the remaining 15 % by secondary crops was assigned to the main crops. An 'average' typical monthly crop calendar was assigned to each zone, based on work done by FAO's global information and early warning system, and on information from the reference library of FAO's agro-meteorology group, AQUASTAT and, for eastern Africa, from the IGADD crop production system zones inventory.

For each crop the actual cropping intensity was derived from national crop production and land use figures extracted from the FAO AGROSTAT and AQUASTAT databases. It ranges from 100 to 200%, according to the crop calendar. The cropping intensity to be used in this study of irrigation potential ('potential' scenario) was generally estimated by increasing current values by 10 to 20%, but it was assumed that because of market limitations the current high intensity (in relative terms) of vegetables in certain parts of the continent would not be found in the potential scenario. Therefore, intensities of cereal crops are higher in the potential scenario than in the actual situation.

Definition of the Climate Stations' area of Influence

The climate data from the FAOCLIM cd-rom were used, as this was the most up to date climate database available. This data set includes long term average rainfall and reference potential evapotranspiration (ET_0) data for 1025 stations throughout Africa. ET_0 was calculated by the Penman-Monteith method.

To obtain a spatial coverage of climate data (P. ET_0) over the continent, each station was assigned an area of influence using the Thiessen polygons method. This method assigns an area of 'nearest vicinity' to each climate station. Gives an indication of the density of the stations over the continent. As expected, the desert areas in northern and southern Africa are much less well covered than the rest of the continent. The rainfall data were compared with raster maps prepared by the Australian National University and corrected where necessary.

Combination of Cropping Pattern Zones with the Climate Stations

In ArcInfo, the 24 cropping pattern zones and the 1025 climate station data were merged. This resulted in 1437 basic map features, homogeneous in irrigation cropping characteristics and climate. All further calculations were carried out on these 1437 basic mapping units.

Results

Summarizes the figures for each of the 84 zones. NIWR and GIWR for the potential scenario with effective rainfall were further combined with the 136 basic units of this study to obtain individual NIWR and GIWR for each of these units through GIS. The results have been compared with figures available from country studies (national water master plans, projects, etc.). The comparison shows that the methodology yields relatively accurate regional estimates of IWR that are suitable for the present study. Discrepancies with country studies find their origin mostly in the assumptions made on cropping pattern, cropping intensity and irrigation efficiency.

The influence of *cropping pattern zones* on the quality of the output is of prime importance. Important differences in irrigation water requirements in adjacent zones is one of the consequences of this approach. For instance, in Burkina Faso, areas located north of the 1000 mm annual rainfall line have a gross potential water requirement of 500 mm per year, while areas located just south of this line need more than 2800 mm per year. This artificial break is due to the choice of the irrigated cropping pattern zones, where it was decided that no rice was cultivated under 1000 mm of rainfall per year. Within the cropping pattern zones, the boundaries of irrigation water requirement zones follow rainfall trends.

The differences in irrigation water requirements between adjacent zones is directly related to the *density of the climate stations' network*. In low-density areas, such as the Sahara and southern Africa, differences in IWR between adjacent zones are high (up to 600 mm/ year gross requirements) as the low station density does not allow the delineation of HIWR zones with smaller differences. A high density of the station network in the rest of the continent, in combination with rainfall raster maps, has resulted in differences of a maximum of 200 mm/year gross requirements between adjacent zones.

Irrigation Cropping Pattern Zones. List of Cropping Pattern Zones.

1. Mediterranean coastal zone
2. Saharan oases
3a Semi-arid to arid savannas in West-East Africa
3b Semi-arid/arid savanna (Somalia-Kenya-Southern Sudan)
4a Rice: Niger/Senegal rivers
4b Rice: Gulf of Guinea
4c Rice: Southern Sudan
4d Rice: Madagascar tropical lowland
4e Rice: Madagascar highland
5 Egyptian Nile and delta
6 Ethiopian highlands
7 Sudanese Nile area
8 Shebelli-Juba river area in Somalia
9 Rwanda-Burundi-Southern Uganda highlands
10 Southern Kenya-Northern Tanzania
11 Malawi-Mozambique-Southern Tanzania
12a West and Central African humid areas above the equator
12b Central African areas below the equator
13 River effluents on Angola-Namibia-Botswana border
14 South Africa-Namibia-Botswana desert and steppe
15 Zimbabwe highland
16 South Africa-Lesotho-Swaziland
17 Awash river area in Ethiopia
18 All islands (Comoros-Mauritius-Seychelles-Cape Verde).

Environmental Considerations in Irrigation Development

Irrigation has contributed significantly to poverty alleviation, food security, and improving the quality of life for rural populations. However, the sustainability of irrigated agriculture is being questioned, both economically and environmentally. The increased dependence on irrigation has not been without its negative environmental effects.

Inadequate attention to factors other than the technical engineering and projected economic implications of large-scale irrigation or drainage schemes in Africa has all too frequently led to great difficulties. Decisions to embark on these costly projects have often been made in the absence of sound objective assessments of their environmental and social implications. Major capital intensive water engineering schemes have been proposed without a proper evaluation of their environmental impact and without realistic assessments of the true costs and benefits that are likely to result.

The sustainability of irrigation projects depends on the taking into consideration of environmental effects as well as on the availability of funds for the maintenance of the implemented schemes. Negative environmental impacts could have a serious effect on the investments in the irrigation sector. Adequate maintenance funds should be provided to the implementing organizations to carry out both regular and emergency maintenance.

It is essential that irrigation projects be planned and managed in the context of overall river basin and regional development plans, including both the upland catchment areas and the catchment areas downstream.

Potential Environmental Impacts of Irrigation Development

The expansion and intensification of agriculture made possible by irrigation has the potential for causing: increased erosion; pollution of surface water and groundwater from agricultural biocides; deterioration of water quality; increased nutrient levels in the irrigation and drainage water resulting in algal blooms, proliferation of aquatic weeds and eutrophication in irrigation canals and downstream waterways. Poor water quality below an irrigation project may render the water unfit for other users, harm aquatic species and, because of high nutrient content, result in aquatic weed growth that obstructs waterways and has health, navigation and ecological consequences. Elimination of dry season die-back and the creation of a more humid microclimate may result in an increase of agricultural pests an plant diseases. Large irrigation projects which impound or divert river water have the potential to cause major environmental disturbances, resulting from changes in the hydrology and limnology of river basins. Reducing the river flow changes flood plain land use and ecology and can cause salt water intrusion in the river and into the groundwater of adjacent lands. Diversion of water through irrigation further reduces the water supply for downstream users, including municipalities, industries and agriculture. A reduction in river base flow also decreases the dilution of municipal and industrial wastes added downstream, posing pollution and health hazards.

The potential direct negative environmental impacts of the use of groundwater for irrigation arise from over-extraction (withdrawing water in excess of the recharge rate). This can result in the lowering of the water table, land subsidence, decreased water quality and saltwater intrusion in coastal areas. Upstream land uses affect the quality of water entering the irrigation area, particularly the sediment content (for example from agriculture-induced erosion) and chemical composition (for example from agricultural and industrial pollutants). Use of river water with a large sediment load may result in canal clogging. The potential negative environmental impacts of most large irrigation projects described more in detail below include: waterlogging and salinization of soils, increased incidence of water-borne and water-related diseases, possible negative impacts of dams and reservoirs, problems of resettlement or changes in the lifestyle of local populations.

Waterlogging and Salinization

About 2 to 3 million ha are going out of production worldwide each year due to salinity problems. On irrigated land salinization is the major cause of land being lost to production and is one of the most prolific adverse environmental impacts associated with irrigation. However, very limited research has yet been conducted to quantify the economic impact of irrigation induced salinization. Quantitative measurements have generally been limited to the amount of land affected or abandoned. Estimates of the area affected have ranged from 10 to 48% of worldwide total irrigated area. Especially the arid and semi-arid areas have extensive salinity problems.

Waterlogging and salinization of soils are common problems associated with surface irrigation. Waterlogging results primarily from inadequate drainage and over-irrigation and, to a lesser extent, from seepage from canals and ditches. Waterlogging concentrates salts, drawn up from lower in the soil profile, in the plants' rooting zone. Alkalization, the build-up of sodium in soils, is a particularly detrimental form of salinization which is difficult to rectify. Irrigation-induced salinity can arise as a result of the use of any irrigation water, irrigation of saline soils, and rising levels of saline groundwater combined with inadequate leaching. When surface water or groundwater containing mineral salts is used for irrigating crops, salts are carried out into the root zone. In the process of evapotranspiration, the salt is left behind in the soil, since the amount taken up by plants and removed at harvest is quite negligible. The more arid the region, the larger is the quantity of irrigation water and, consequently, the salts applied, and the smaller is the quantity of rainfall that is available to leach away the accumulating salts. Excess salinity within the root zone reduces plant growth due to increasing energy that the plant must expend to acquire water from the soil. The tolerance of crops to salinity is variable: clover and rice are more sensitive to salts than barley and wheat. Comprehensive studies of farm-level effects of irrigation-induced salinity indicate that the yields of paddy and wheat are around 50% lower on the degraded soils and net incomes in salt-affected lands are around 85% lower than the unaffected land.

Irrigation-related salinity has adverse effects not only on the production areas, but also on areas and people downstream. The rivers, particularly in arid zones tend to become progressively more saline from their headwaters to their mouths. The aquifers interrelated with the river are highly saline and the salts discharged to the river system from saline aquifers adversely affect downstream water users, particularly irrigated agriculture and, in some special cases, wildlife. Many of the soil-related problems could be minimised by installing adequate drainage systems. In Egypt, for example, the installation of drainage systems effectively reduced soil salinity. The average yield for wheat increased from 1 ton/ha before drainage to about 2.4 tons/ha. Similarly, the yield for maize increased from 2.4 tons/ha to 3.6 tons/ha after drainage infrastructure was completed. Drainage is a critical element of irrigation projects, that however still too often is poorly planned and managed. Waterlogging can also be reduced or minimized, in some cases, by using micro-irrigation which applies water more precisely and can more easily limit quantities to no more than the crops needs.

Water-borne and Water-related Diseases

Water-borne or water-related diseases are commonly associated with the introduction of irrigation. The diseases most directly linked with irrigation are malaria, bilharzia (schistosomiasis) and river blindness (onchocerciasis), whose vectors proliferate in the irrigation waters. Other irrigation-related health risks include those associated with increased use of agrochemicals, deterioration of water quality, and increased population pressure in the area. The reuse of wastewater for irrigation has the potential, depending on the extent of treatment, of transmitting communicable diseases. The population groups at risk include agricultural workers, consumers of crops and meat from the wastewater-irrigated fields, and people living nearby. Sprinkler irrigation poses an additional risk through the potential dispersal of pathogens through the air.

The risk that one or more of the above diseases is introduced or has an increased impact is most likely in irrigation schemes where:

- soil drainage is poor, drainage canals are either absent, badly designed and/or maintained;
- rice or sugar cane is cultivated;
- night storage reservoirs are constructed;
- borrow pits are left with stagnant water;
- canals are unlined and have unchecked vegetation growth.

Malaria

Malaria is by far the most important disease, both in terms of the number of people annually infected, and whose quality of life and working capacity are reduced, and in terms of deaths. Worldwide, some 2000 million people live in areas where they are at risk of contracting malaria. The total number of people infected with malaria is estimated at 100 to 200 million with between 1 and 2 million deaths per year, with almost 90% of the cases in Africa. Drug treatment has become difficult recently because the parasite has become resistant to certain drugs that have been used for a long time in many parts of the world. Interruption of disease transmission chemicals for the control of the vector, the mosquito, has become less effective because some mosquito vector species have become resistant to previously effective insecticides and some insecticides have been banned for environmental reasons.

Bilharzia

Bilharzia is almost as widespread as malaria, but rarely causes immediate death. An estimated 200 million people are infected and the transmission occurs in some 74 countries. The infection is particularly common in children who play in water inhabited by the snail intermediate host.

Severe infection in childhood leads to long-term damage to bladder, kidneys and liver, which may cause death many years after the original infection. Infection at any age may make people feel unwell and reduce working capacity.

Bilharzia is an infection caused by parasitic worms or blood flukes of certain species of the genus *Schistosoma*. Adult parasites live in the blood of mammals, but their life cycle requires a phase of asexual multiplication within a freshwater snail host. The flukes infect humans who enter their exposed skin in water, usually through swimming, bathing or wading. There exists either urinary or intestinal schistosomiasis. The type and extent of health complications associated with schistosomiasis appear to vary with species and strain of parasite and by the characteristics of the human population.

As shown in some examples below, water resources development projects may make schistosomiasis worse, and that there are serious public health consequences in many cases. Specifically, WHO stated that in areas endemic for schistosomiasis, water resources development projects should have schistosomiasis prevention and control built into programme design and implementation. Furthermore, even in cases where irrigation does not increase schistosomiasis infection rates, careful studies should be made of snail species and of existing patterns of schistosomiasis transmission. Further, a percentage of investment and operating funds should be allocated for appropriate water supply and sanitation and for health care to treat local populations for any water-related or other ailments associated with the project.

Effects of Large Dams and Reservoirs on the Prevalence of Bilharzia in Africa

In Africa, in recent decades cautionary warnings have accompanied many irrigation and dam projects regarding the likely impact of increased schistosomiasis transmission. In some cases, these prophecies came true, including the Volta Lake project in Ghana in the 1960s and the Kainji Lake project in Nigeria around 1970. Most recently, research appears to confirm the prediction, that constructing a dam across the mouth of the Senegal River would lead to a surge in schistosomiasis transmission.

In part of Upper Egypt, schistosomiasis prevalence is said to have increased from 6 to 60% three years after the Aswan low dam was completed in the early 1930s. Following the construction of the Sennar dam in Sudan in 1926 and the Gezira scheme which followed, schistosomiasis spread. At the Arusha Chini irrigation project in Tanzania, research reported the prevalence to be 85 % in 1962. On the southwestern shore of Lake Volta in Ghana, prevalence is reported to have reached 90% two years after the lake was filled in 1966.

In Zambia, prevalence of schistosomiasis around Lake Kariba in 1968, 10 years after the Zambezi river was dammed, was 15% in adults and 70% in children. At Kainji Lake in Nigeria, prevalence increased from 30% to 45%.

In Ethiopia, the two Koka dams in the Awash Valley present a case of the different effects a dam has on the two different forms of schistosomiasis through the reservoir, downstream hydrology, and irrigation. It was reported that these two dams, built in 1958 and 1964, appeared to have no effect on urinary schistosomiasis transmission through their reservoirs as assessed in 1968. The dams did, however, enable intestinal schistosomiasis transmission to occur in the upper Awash Valley by changing its hydrology and introducing irrigation.

In West Africa, both urinary and intestinal forms of schistosomiasis became highly endemic in the Office du Niger area. The irrigation scheme had a mean prevalence of urinary schistosomiasis of 64.4%, compared to the river communities' 19.9%, and a mean prevalence of intestinal schistosomiasis of 53.9%, compared to the river communities' 1.9%.

Effects of Small Dams on the Prevalence of Bilharzia in Africa

Throughout the semi-arid areas of Africa, people have constructed many small earthen dams to provide irrigation water for dry season cultivation. While reports often blame these dams for spreading schistosomiasis, there is little evidence to substantiate such claims. Small dams in the semi-arid zone of West Africa get the greatest amount of attention. Although the dams built for dry-season irrigation extended the range of *Bulinus rohlfsi* snails, a common vector for urinary schistosomiasis in West Africa, this did not cause any noticeable increases in prevalence of urinary schistosomiasis.

Several studies in the same Sudano-Sahel ecological zone as northern Nigeria noted that evidence linking small earthen dams to schistosomiasis was lacking. In Burkina Faso, natural seasonal ponds infested by *Bulinus truncatus snails* are more common and dangerous locations of infections than are artificial reservoirs created by small dams.

In Cameroon, a nation-wide survey failed to find an association between the rate of schistosomiasis and small dams. Instead, temporary ponds and snail hosts adapted to low seasonal rainfall permits intense transmission of urinary and intestinal schistosomiasis throughout northern Cameroon, regardless of dams. Contrary to the experience in Nigeria, Burkina Faso and Cameroon, one study in northern Ghana showed small dams to be linked to schistosomiasis. Data collected during 1960-61 showed much higher prevalence of schistosomiasis in the eastern, more densely settled part of what was the Upper region of Ghana. The mean prevalence of urinary schistosomiasis in the region's eastern part was 19.8% in 15 districts without dams, 42.3% in 16 districts with dams 12 years old, and 52.0% for 6 districts with dams 3 years old. Areas in the western part of the region with few or no dams had infection rates under 10% in 6 districts, and 10 to 29% in 10 others. In contrast, prevalence was over 70% in 2 of the 3 western districts containing dams.

The Control of the Water-related Diseases

The control of the water-related diseases can be effected in a number of ways, some of which are mutually reinforcing. Three types of measures are distinguished:

- measures aimed at the pathogens: immunization, prophylactic or curative drugs;
- measures aimed at reducing vector densities or vector lifespan: chemical, biological and environmental controls;
- measures to reduce human/vector or human/pathogen contact: health education, personal protection measures and mosquito proofing of houses.

Of the above, environmental control measures are considered to be long-lasting and environmentally-sound. These include preventing or removing aquatic vegetation, lining canals with cement or plastic, regularly fluctuating water levels, periodic rapid drying of irrigation canals, preventing contamination of water bodies with faeces, supply of safe and clean drinking water, appropriate siting of housing of the farmers etc. For example, in Zimbabwe, in a communal small-holder irrigation project at Mushandike, adoption of a these measures resulted within three years in a drop of the infection rate from an initial 70 to 80% to virtually nil.

Potential Environmental Impacts of Dams and Reservoirs

The benefits of a dam project are flood control and the provision of a more reliable and higher quality water supply for irrigation, domestic and industrial use. Intensification of agriculture locally through irrigation can reduce pressure on uncleared forest lands, intact wildlife habitat and marginal agricultural land. In addition, dams create reservoir fishery and the possibilities for agricultural production on the reservoir drawdown area, which more than compensate for losses in these sectors due to the dam construction.

However, large dam projects cause irreversible environmental changes over a wide geographic area and thus have the potential for significant impacts. Criticism of such projects has grown in the last decade. Severe critics claim that because benefits from dams are outweighed by their social, environmental and economic costs, the construction of large dams is unjustifiable. In some cases, environmental and social costs can be avoided or reduced to an acceptable level by carefully assessing potential problems and implementing cost-effective corrective measures.

Damming the river and creating a lake-like environment profoundly changes the hydrology and limnology of the river system. Dramatic changes occur in the timing of flow, quality, quantity and use of water, aquatic biota, and sedimentation in the river basin. The area of influence of a dam project extends from the upper limits of the catchment of the reservoir to as far downstream as the estuary, coast and offshore zone. While there are direct environmental impacts associated with the construction of the dam (for example dust, erosion, borrow and disposal problems), the greatest impacts result from the impoundment of water, flooding of land to form the reservoir and alteration of water flow downstream. These effects have direct impacts on soils, vegetation, wildlife and wildlands, fisheries, climate and especially the human populations in the area.

Increased pressure on upland areas above the dam is a common phenomenon caused by the resettlement of people from the inundated areas and by the uncontrolled influx of newcomers into the basin catchment. On-site environmental deterioration as well as a decrease in water quality and increase in sedimentation rates in the reservoir result from clearing of forest land for agriculture, grazing pressures, use of agricultural chemicals, and tree cutting for timber or fuelwood.

The Impact of Dams on Flooding

The function of dams and reservoirs in flood control is to reduce the peak flows entering a flood prone area. Rather than maintaining high water levels for increased head or sustained water supply for irrigation, flood control operation requires that water levels be kept drawn down deliberately prior to and during the flood season in order to maintain the capacity to store any incoming floodwater. However, flood plains may be productive environments because flooding makes them so. Flooding recharges soil moisture and replenishes the rich alluvial soils with flood deposits of silt. In arid areas flooding may be the only source of natural irrigation and soil enrichment. Reduction or elimination of flooding has the potential for impoverishing flood recession cropping, groundwater recharge, natural vegetation, wildlife and livestock population in the flood plain which are adapted to the natural flood cycles.

To maintain the productivity level of the natural systems, compensatory measures have to be taken, such as fertilization or irrigation of agricultural lands. In addition, when channelization measures reduce the frequency of flooding, the sediments entering the river systems from catchment areas upstream will be passed to the mouth of the river unless overflow areas are present downstream. Channel modification can result in a number of negative environmental impacts. Any measure that increases the velocity of flow increases the erosive capacity of the water. Although channel improvement can alleviate flooding problems in the treatment area, flood peaks are likely to increase downstream, thus simply transferring the problem elsewhere. Dikes built on the flood plain to exclude water from certain areas affect the hydrology of the area, and can have impacts on wildlife and livestock habitat and movement.

The Impact of Dams on Fisheries and Wildlife

Fishery alongside the rivers usually declines due to changes in river flow, deterioration of water quality, water temperature changes, loss of spawning grounds and barriers to fish migration. A reservoir fishery, sometimes snore productive than the previous fishery alongside the river, however, is created.

In rivers with biologically productive estuaries, both marine and estuarine fish and shellfish suffer from changes in water flow and quality. Changes in freshwater flows and thus the salinity balance in an estuary will alter species distribution and breeding patterns of fish. Changes in nutrient levels and a decrease in the quality of the river water can also have profound impacts on the productivity of an estuary. These changes can also have major effects on marine species which feed or spend part of their life cycle in the estuary, or are influenced by water quality changes in the coastal areas. The greatest impact on wildlife will come from loss of habitat resulting from reservoir filling and land use changes in the catchment area. Migratory patterns of wildlife may be disrupted by the reservoir and associated developments. Aquatic fauna, including waterfowl, amphibians and reptiles can increase because of the reservoir.

Socio-economic Impacts Irrigation Schemes

The objective of irrigation projects is to increase agricultural production and consequently to improve the economic and social well-being of the rural population. However, changing land use patterns may have other impacts on social and economic structure of the project area. Small plots, communal land use rights, and conflicting traditional and legal land rights all create difficulties when land is converted to irrigated agriculture. Land tenure/ownership patterns are almost certain to be disrupted by major rehabilitation works as well as a new irrigation project. Similar problems arise as a result of changes to rights to water. Increased inequity in opportunity often results from changing land use or water use patterns. For example, owners benefit in a greater proportion than tenants or those with communal rights to land. Access improvements and changes to the infrastructure are likely to require some field layout changes and a loss of some cultivated land.

Irrigation projects tend to encourage population densities to increase, either because of the increased production of the area or because they are part of a resettlement project. Impacts resulting from changes to the demographic/ethnic composition may be important and have to be considered at the project planning stage through, for example, sufficient infrastructure provision.

The most significant issue arising from large dam construction is resettlement of people displaced by the flooding of land and homes. This can be particularly disruptive to communities and insensitive project development would cause unnecessary problems by lack of inadequate compensation of the affected population. Human migration and displacement are commensurate with a breakdown in community infrastructure which results in a degree of social unrest and may contribute to malnutrition. As an example of the number of people displaced by the construction of a dam, filling of the reservoir behind the High Aswan Dam displaced 50000 to 60000 people in Egypt and some 53000 people in the Sudanese portion.

Changing land patterns and work loads resulting from the introduction or formalizing of irrigation are likely to affect men and women, ethnic groups and social classes unequally. Groups that use common land to make their living or fulfil their household duties, for example for charcoal making, hunting, grazing, collecting fuel wood, growing vegetables, etc. may be disadvantaged if that same land is taken over for irrigated agriculture or for building irrigation infrastructure. Women, migrants groups and poorer social classes have often lost access to resources and gained increased work loads. Conversely, the increased income and improved nutrition from irrigated agriculture may benefit women and children in particular. The most common socio-economic problems reducing the income generating capacity of irrigation schemes are:

- The social organization of irrigation operation and maintenance (O&M). Poor O&M contributes significantly to long-term salinity and waterlogging problems and needs to be adequately planned at the design stage to sustain the long-term development of the schemes.
- Reduced farming flexibility. Irrigation may only be viable with high-value crops, thus reducing extensive activities such as grazing animals, operating woodlots, etc.
- Changing labour patterns that make labour-intensive irrigation unattractive.
- Insufficient external supports such as markets, agrochemical inputs, extension and credit facilities.

User participation at the planning and design stages of both new schemes and the rehabilitation of existing schemes, as well as the provision of extension, marketing and credit services, can minimize negative impacts and maximize positive ones.

Alternatives to Mitigate the Negative Impacts of Irrigation Projects

Alternatives exist to mitigate adverse effects of irrigation development. Some of them are listed below:

- locating the irrigation project on the site where negative impacts are minimized;
- improving the efficiency of existing projects and restoring degraded croplands to use rather than establishing a new irrigation project;
- developing small-scale, individually-owned irrigation systems as an alternative to large-scale, publicly-owned and managed schemes;
- using sprinkler irrigation and micro-irrigation systems to decrease the risk of waterlogging, erosion and inefficient water use;

- using treated wastewater, where appropriate, to make more water available to other users;
- maintaining flood flows downstream of the dams to ensure that an adequate area is flooded each year, among other reasons, for fishery activities.

The Role of Wetland and the Impacts of Water Development Projects

Wetlands are wildlands of particular importance both economically and environmentally. The most important roles which wetlands perform are:

- Preservation of biological diversity: for many species of shrimp, fish and waterfowl, tidal and fresh water marshes, coastal lagoons and estuaries are of vital importance as breeding grounds as well as staging areas in their migration routes. All types of wetlands may harbour unique plants and animals.
- Production of goods: wetlands are among the most productive ecosystems in the world. Estuaries and tidal wetlands, in particular mangroves, are important nursery areas for most species of fish and shrimp which are later caught offshore. Shallow water areas are, in general, rich fishing grounds. Flood plains are important grazing areas for cattle and wildlife and vital spawning grounds for many fish species. Swamp forest may yield valuable timber.
- Production of services: wetlands can contribute to local rainfall and can be an efficient, low cost water purification system (herbaceous swamps), a recreation area (hunting, fishing, boating), a buffer against floods, and provide protection against coastal erosion by storms (mangroves).

Despite their importance, wetlands are under threat, in particular, from direct conversion of wetlands for agriculture and projects which affect the hydrology of a wetland, such as construction of dams, flood control, lowering of the aquifer drainage, and irrigation and other water supply systems.

The sections below describe some important wetlands in Africa, but the list is far from exhaustive.

Wetlands in the West African Sahel

The Senegal river waters the west of Mali, the north of Guinea, the north of Senegal and the south of Mauritania, the main Niger river stretches through Mali, the Republic of Niger and Nigeria. The frontiers of Niger, Nigeria, Cameroon and Chad meet in the Lake Chad. For all of these countries, river valleys and lakes are of the utmost economic importance because of their high productivity. Mauritania, Mali, Niger and Chad, whose territories are for the most part desert, draw their agricultural resources and the greater part of their animal protein from fishing and stock-raising in the river valleys. The states along the water courses are acting in concert to increase the use of water resources and avoid the hazards of drought. It is a question of ensuring mastery over the flood waters to grow rice and other cereals, and of using surface water for irrigation. These operations inevitably tend to modify ecological conditions. In practice this will lead to significant losses of aquatic habitats and to a noticeable decrease in fish and waterfowl.

Regions such as the southern part of the Sahel, with a strongly seasonal rainfall regime and yet with sufficient rainfall to support seasonal agriculture and pastoralism, support large numbers of people. Flood plains, swamps and lakes provide a range of ecological resources and economic opportunities. Without wetlands, the drylands of the West African Sahel would be both less productive and more hazardous as a place for people to live.

In the semi-arid zone of Western Africa, patterns of rainfall and river flow are strongly seasonal. In northern Nigeria, most of the rainfall occurs in just three or four months, between June and September. During this short rainy season, precipitation exceeds evapotranspiration and runoff occurs. Savannah rivers run strongly, but start to shrink as the rains end. During the wet season rainfed agriculture is possible, and there are extensive grasslands providing relatively nutritious grazing for livestock away from the river valleys. Once the rains end, these resources also dry up, and pastoralists concentrate on the remaining wetlands in river valleys or larger wetlands such as the delta of the Senegal River the Niger Inner Delta in Mali or Lake Chad.

It is also only in these areas that agriculture can continue into the dry season. In many areas rice is planted, either in the rains before the floods arrive, or as the floods recede. Flood-recession crops such as sorghum or beans are planted as the waters recede, and farmers dig wells or use remaining pools to irrigate small gardens using buckets, shadoofs or, more recently, small pumps. At the same time, other economic activities such as fishing are also linked to the changing flood. Many fishes move laterally out of the riverbed pools into the flood plain to breed, and their offspring is caught as the water retreats. Later in the dry season, the residual pools in the riverbed are themselves fished. The high fisheries productivity of most of the seasonally inundated flood plains is fostered, at least in part, by the nutrient-rich dung left by the grazing animals during the previous dry season.

In valleys such as the Senegal or the vast flooded plains of the Niger Inner Delta, the annual cycle of farmers, herders and fishers is closely linked to the seasonal cycle of flooding. The flood plain of the Senegal stretches up to 30 km in width, and runs 600 km downstream of Bakel. It covers a total of about 1 million ha and supports farmers, pastoralists and fishing communities. Up to half a million people depend on the flood-related cropping in the 'waalo' land of the flood plain.

There is growing evidence that large-scale capital intensive water development schemes do provide neither the range of foodstuffs nor the economic return of traditional systems. Studies, comparing the efficiency per unit of water of traditional extensive systems of cultivation, grazing and fishing in the Niger Inner Delta with the intensive modern project of the Office du Niger, showed that both systems produce about the same gross profit margin, even when the running costs and management charges for the irrigation scheme are taken into account. However, the extensive system produces meat, milk, fish and rice compared to the rice-only irrigated system. More importantly, when the interest charges arising from the irrigation scheme are taken into account, the net profit from the irrigated rice turns into a loss of $0.65/100 m³ of water, whilst the extensive traditional methods benefit from the 'free services of nature' and turn in a net profit of $0.42/100 m³ at 1984/85 prices.

The Hadejia-Nguru Wetlands

The Hadejia-Nguru Wetlands concern a part of the flood plain of the Komadougou-Yobe river basin in the Lake Chad basin in the north-east of Nigeria and are home to probably about a million people. The wetlands have formed where the waters of the Hadejia and Jama'are rivers meet the lines of ancient sand dunes aligned northeast-southwest. An area of confused drainage has formed here, with multiple river channels and a complex pattern of permanently and seasonally flooded land and dryland. The wetlands are nationally and internationally important for migratory waterfowl. The wetlands support extensive wet-season rice farming, flood-recession agriculture and dry-season irrigation. The flood plain also supports large numbers of fishing people, most of whom also farm, and is grazed by very substantial numbers of Fulani livestock, particularly cattle, which are brought in from both north and south in the dry season. There is also an important dispatch from the wetlands of fuelwood and fodder for horses. In the past, much of the rice, as well as fish and birds, was traded out of the area. This has changed, but there is now a strong export of other agricultural products, for example peppers, wheat and fuelwood. The economic value of production from the wetlands is very large, many times greater than that of all the irrigation schemes for which the inflowing rivers are dammed, diverted and their waters used.

There are natural changes, for example the impacts of drought, that have serious implications for the future of the wetlands and the sustainability of their production systems. There are also major economic changes within the wetlands themselves. The extent of irrigation has greatly increased over the 1980s, largely as a result of the advent of small petrol-powered pumps and the ban on the importation of wheat in 1988. As the use of small pumps spreads, conflicts are beginning to emerge between farmers and pastoralists, and between small and large farmers for access to land.

The wetlands have also been affected by developments elsewhere in the river basin. The construction of the Tiga Dam on a tributary of the Hadejia river in the early years of the 1970s has exacerbated the effects of the low rainfall of the last two decades. The result has been a reduction in the extent of flooding in the wetland. Construction of a dam on the Hadejia river just above Hadejia town to provide short-term storage of water to irrigate the Hadejia Valley Project Phase 1 began, in the early 1980s, but was stopped for several years because of financial problems. The main dam was completed in 1992, soon after work restarted on the project. The dam has created a large shallow lake upstream and it will probably have a major effect on the timing and extent of flooding in the wetlands.

Most of the dams, irrigation schemes and water resources plans for the Yobe basin were prepared in the 1970s and early 1980s, using data for the relatively wet period up to 1973. The post-1972 drought has reduced the proportion of rainfall which runs off to the rivers. The 1988 flood at Hadejia was probably one of the largest for some years and it was augmented by the failure of the dam at Bagauda.

The Hadejia-Nguru wetlands have long been known as a centre of fish production. Upstream hydrological developments induced by irrigation projects threaten to degrade this important resource. Studies of flood plain fisheries have shown that fish production is closely related to flood extent. The existing and planned dams upstream of the Hadejia-Nguru wetlands are likely to have a serious impact on fisheries. Despite the lack of information specific to the Hadejia-Nguru wetlands, there are enough studies from other flood plains affected by hydraulic works to show that the effects of dams on fish communities are likely to be serious. The dams are likely to bring changes in river flow, loss of habitat, blocking of channels, changes in silt loading, plankton abundance and temperature which are likely to affect fish communities.

The economic value of fish production from the flood plains adds weight to the argument in favour of maintaining the annual flooding of the wetlands. Moreover, the significance of fishing goes beyond its value in monetary terms. Fishing plays an important role in the flexibility and adaptability of the rural economy in the flood plains. A reduction in this flexibility through degradation of the fishery resource may have serious repercussions on the ability of communities to adapt to fluctuations in their environment. Many people are involved in the fisheries and so the social consequences of any appreciable reduction in productivity will be felt throughout the area. Degradation of the fisheries may also affect other sectors of the rural economy. Most people who fish also pursue other activities-such as farming, livestock rearing, manufacturing of crafts or trading-and the loss of, or reduction in one component of the household economy is likely to affect activities in other sectors. There will also be 'downline' effects on fish processors, fish dealers, customers and consumers.

In addition to producing fuelwood, the forest reserves and bushland of the flood plains yield important non-timber forest products that are significant to the livelihoods and subsistence of local communities. Some, including leaves, are important marketed commodities that generate substantial income. *Doum* palm leaves are either processed into mats and other products or sold as raw material. The harvesting and processing of doum palm leaves is a dry season activity, and many people migrate to the wetlands to harvest the palm. Mat-making from doum is also a specialized activity of many flood plain villages. Mats and other doum products, for example rope and baskets, are sold locally or exported to other regions. *Baobab* leaves are used widely as an ingredient for soups and stews and are especially important as a 'drought food'. Honey, produced by local beekeepers, is a highly valued commodity.

Since 1985, the area has been the focus of the Hadejia-Nguru Wetlands Conservation Project. This project has been run jointly by the Nigerian Conservation Foundation, IUCN (International Union for the Conservation of Nature), the Royal Society for the Protection of Birds and the International Council for Bird Preservation (now renamed Birdlife International). In 1990 a major development project was started by the European Community that included the eastern part of the area. The North East Arid Zone Development Project (NEAZDP) has a very substantial budget to generate village-based development initiatives. Attention has tended to be directed in particular to the potential resources of the wetlands.

Wise use of the wetlands of the Hadejia-Nguru wetlands demands a proper understanding of the environmental and socio-economic changes that are occurring and of those that may be predicted. Understanding of the impacts of changes inside and outside the flood plain is far from easy, and prediction of future impacts is even harder. However, without such understanding and prediction, effective planning and management is impossible.

The economic importance of the flood plains suggests that benefits it provides cannot be excluded as an opportunity cost of any scheme that diverts water away from the flood plain system. Policy makers should be aware of this problem when designing water development projects in the river system. Further analysis is also required of the type of 'regulated flood projects' regime, which could maintain much of the flood plain system intact while still allowing some upstream water developments. Further investigation of all the economic benefits provided by the wetlands is also needed, and the sustainability of production within a flood plain area should be more thoroughly examined.

Effects of the Jonglei Canal on the Sudd Swamps

In the southern of Sudan, the Nile discharges its water into the great wetlands of the Sudd, a network of channels, lakes and swamps in which as much as half of the inflowing water is disappears through evaporation. The Jonglei Canal was designed to bypass the Sudd and direct downstream a proportion of the water that is 'lost' from the Nile each year by spill and evaporation in the swamps. The projected dimensions of the canal are as follows: a width from bank to bank of about 75 metres, a channel bed-width averaging 38 metres, a depth varying from 4 to 8 meters, and a length of 360 km, over twice the length of the Suez canal. Jonglei is a small Dinka village close to the

Atem channel at a point where the canal alignment was planned to begin. Although the offtake will now be further south at Bor, the canal is still so named and Jonglei has given its name to a province as well.

The canal has not been completed, but detailed surveys were undertaken to determine a whole range of effects, many of which will be shown to be disadvantageous to the inhabitants of the Jonglei Area. Some of the effects are described below.

The river-flooded grasslands are an essential seasonal resource during the driest months of the year. Not only is there drinking water available in the rivers, but the process of seasonal inundation itself produces species of grasses which sustain the herds from about January until April. There are no other alternatives as the grasses of the high land are exhausted or reserved for the livestock (mainly smaller stock), held by the few people who elect to remain behind, and the rain-flooded grasslands have become woody and unpalatable and produce little or no regrowth after burning. It follows that the river-flooded grasslands are crucial to the pastoral economy at this time of the year. It is, however, just these grasslands that may be reduced by the operation of the canal.

The water benefit of the canal downstream will be around 4 km³/year and according to some estimates even an extra water flow of up to 10 km³/year may be reached. These quantities are a substantial percentage of the average 'losses' by the evapotranspiration, the natural production of river-flooded grasses being a function of the annual fluctuation in river discharge and thus of the annual variation in area flooded. In other words, to the local inhabitants these are not losses in water at all, though the waters are excessive and the cause of damaging floods, as in 1964.

The floods of the 1960s, reaching a peak in 1964, caused great damage to human interests. On the Zeraf island alone it was reckoned that 130000 cattle were lost owing to exposure and lack of grazing since practically the whole area remained under water for a long period. Similar disastrous effects occurred west of the Bahr el Jebel in the vicinity of Adok. It follows that any reduction in peak flows could be protective and beneficial. The same model can be applied to give some indication of the effect of the canal on areas of flooding. The figure of 25 million m³/day for a canal diversion may reduce the area of flooding by about 19% at a 1964 peak discharge.

The established fisheries of some large lakes in the Sudd are said to have been adversely affected by increased water depth, but, overall, the flooding of the 1960s has multiplied the number of perennial lakes in the system and, thereby, the fishing potential. A severe decrease in the discharge into the Sudd resulting from the Jonglei canal would bring about the total disappearance of many lakes in the papyrus zone and reduce others to the status of seasonal lagoons, with a serious loss of year-round fish and fishing potential. There is, however, likely to be considerable disadvantage to the people of the Zeraf Island and those living west of the Bahr el Jebel, in that mainstream traffic will follow the canal and the old western landing places will be ill-served. In the past, moreover, river traffic has been a major factor in keeping the channels open. Oil prospecting is likely to restart once peace has been restored and this may mean that the companies concerned will wish to keep channels clear. However, if discharges drop to the low figures prior to 1961, the canal could become too shallow for commercial traffic and for the movement of fisheries barges.

The canal will in many areas drive a barrier between wet season villages and dry season grazing grounds along the river channels and therefore dislocate the pastoral cycle. Many people living east of the canal will have to cross it with their livestock when regrowth from rain-flooded grasslands is exhausted and they have to move westwards to the river-flooded grasslands of the Nile. Reinforced structures at various points along the canal are needed to facilitate the crossing of livestock without damage to the embankments and to provide suitably designed boats more efficient than the usual 'dug-out' canoe. Crossing the canal will present a massive logistical problem and besides, raises questions of land ownership among those who may need to cross the canal and cross each others' territory in order to do so. There exists a kind of 'Jonglei Controversy'. The criticism of the environmentalists are many but can be segregated into charges that the Jonglei Canal will drastically affect climate, groundwater recharges, silt and water quality, the destruction of fish and changes in the lifestyle of the Nilotic people. However, other studies claim that the positive effects will counterbalance by far the negative effects. As is the case with the Hadejia-Nguru wetlands understanding and prediction of the impacts is very difficult. However, without such understanding and prediction, effective planning and management is impossible.

Regional Aspects of Environmental Impacts and 'Hot Spots'

This section summarizes the regional outlook for the main African sub-regions with regard to the impact of irrigation on the environment. For each of the sub-regions, environmentally salient features, particularly in relation to irrigation development issues, are presented. Wherever possible, environmental 'hot spots' are identified and described.

The Arid North African Sub-region

The North African sub-region lies in arid or semi-arid zones where the water resources are minimal and where evaporation and seepage losses are very large. The sub-region includes two main zones: the Nile Basin in the east and the western part (Morocco, Algeria and Tunisia). This ecological zone is fragile and agricultural production is regulated by alternating periods of water surpluses and deficiencies. Irrigation is the main alternative to cover the food requirements of the increasing population per unit of agricultural land.

The 'hot spots' of the sub-region are the main rivers located in the arid zones threatened by the irrigation-induced salinization of the soils and more generally the degradation of irrigated lands resulting from poor irrigation management and practices. Field drainage and removal of drained water from the irrigated zone is necessary to limit the risk of soil degradation and salinization. Establishing field drainage is costly, as is the provision of a main drainage network. Moreover, disposal of drainage water represents a major problem. The concentration of salt increases gradually from upstream to downstream as a result of the drainage water inflows.

In the Mediterranean coastal zone, reduction of flow systematically induces sea water intrusion problems.

The Sudano-sahelian Belt

The natural capital of this ecological zone is the most fragile, evincing most of the negative effects of irrigation projects on the environment due to the poor soils, extremely variable rainfall and high risk of drought. Soil degradation has substantially increased the risk of decertification because of mutually reinforcing factors including: loss of organic matter and nutrients, soil structure deterioration and surface crusting, which in turn decreases water infiltration and retention, aggravated by the irrigation-induced salinization.

The environmental degradation has been both a cause and a consequence of poverty, with the Sudano-Sahelian belt comprising some of the poorest countries of the world.

Sudano-Sahelian societies face a formidable challenge which makes the whole belt an environmental 'hot spot'. Pressure is likely to be high on the river valleys and major wetlands such as the Niger Inner

Delta in Mali, the wetland and flood plains in the Lake Chad basin (Hadejia-Nguru, Yaere and the Sudd swamps in the Nile basin in southern Sudan.

Humid West Africa

The natural capital of this ecological zone is relatively favourable in terms of climatic conditions: high and regular rainfall, soils of reasonable quality. However, the high population growth during the last decades has placed the environment under serious stress. About half of the total land area is cultivated under reasonably good conditions with a much lower climatic risk than in the Sahel. Forest land has shrunk to less than a third of the total area, and what remains is decreasing at an alarming rate of 1%, the fastest rate in tropical Africa. The environment of the fragile coastal ecosystems is also threatened by industrial and urban development with increasing pollution levels particularly in the Niger delta of Nigeria. A major part of the biodiversity capital of the sub-region is at risk.

Any upstream irrigation project requires special care in order to avoid negative impacts on the wetlands, mangroves and lagoons located in the coastal zone in the Guinea Gulf, from Guinea Bissau to the Niger delta in Nigeria. This zone is likely to become a continuous urban megalopolis with a population of over 50 million people on 500 km of coastal line. The current development of private small-scale irrigation projects, using groundwater for horticultural crops, could contribute to increase the intrusion of saline waters due to over exploitation of coastal aquifers.

The Congo/Zaire Basin

About three-quarters of the total surface area of the sub-region would theoretically be cultivable but a major part of it is under tropical rainforest. Overall pressure to clear the rainforest is still relatively low, except at the periphery of the sub-region where it interfaces with areas of high population density.

Land currently cultivated represents about 15% of the total area. Agricultural activities are relatively less important in the sub-region compared to the rest of the continent. They are focused on supplying a growing urban market and on permanent plantations.

Irrigated areas are marginal compared to the huge potential of land and surface waters. The irrigation development will have a minimal impact on the environment. Global environmental problems faced by the Congo/Zaire basin are less severe than those of the other sub-regions, although its future development will present a serious challenge. In particular, countries need to preserve the primary rainforest for global biodiversity and climatic reasons.

East Africa

The good soils in the eastern African highlands have favoured the development of intensive agriculture, although soils require conservation measures because of steep slopes. Less favourable lands are cultivated under arid and semi-arid conditions. Forests cover less than 20% of the total area of the sub-region. Due to land scarcity, the primary rainforests with their unique biodiversity are at risk.

Due to the pressure of population on arable lands, the environment is at risk particularly in Kenya with a ratio of 0.2 ha per person. To cover the food deficits, areas under irrigation are increasing. Permanent intensive cropping is the current pattern in favourable highlands, but degradation is high under low-input technology and without adequate erosion control measures. In Tanzania, the central area and the Lake Victoria region represent areas of high population density and areas with a reported high degree of land degradation.

Ethiopia has a very large water resources potential. The development of this resource has been impeded for decades, first by agreements made by colonial powers and shell by political instability. The Ethiopian Blue Nile and other tributaries contribute over 80 % of the water in Sudan and Egypt. The mobilization of this potential would have to take into account environmental and basin issues to mitigate the impact on downstream users.

Southern Africa

The natural capital of the sub-region is very rich in terms of biodiversity and production potential, although large areas are under semi-arid and arid conditions with moderate to high risk of drought.

In some countries, particularly South Africa, past policies have had a negative impact on the environment by encouraging agricultural development through high subsidies on farm inputs and irrigation development without stimulating enough soil and water conservation.

Almost half of the total areas of the sub-region is cultivated with reasonably good soils but climatic conditions are highly variable with a risk of recurrent droughts. To mitigate this risk and to cover the food deficit, areas under irrigation are increasing without significant impact on the environment.

Bibliography

Ashworth, S.: *Seed to Seed*, Decorah, Seed Savers Publications, US, 1991.

Aung, K. Hla, Thomas F. Scherer: *Introduction to Micro-Irrigation*, AE, NY, 2003.

Bander, J.: *Scheduling Irrigation with Evaporation Pans,* Montana State Univ. Agr. Bull. Montana, USA, 1984.

Barghouti, S. & Le Moigne, G.: *Irrigation in Sub-Saharan Africa: Development of Public and Private Systems,* World Bank, Washington, DC, 1990.

Bar-Yosef, B. & Sagiv, B.: *Potassium Symposium*, Pretoria, South Africa, 1985.

Bos, M.G. & Nugteren, J.: *On irrigation Efficiencies*, Wageningen, the Netherlands, Int. Inst. Land Reclam, Improve, 1978.

Bosworth, B.: *Capital Formation and Economic Policy.* Brookings Papers on Economic Activity, 1982

Brooks, D.: *Water: Local-Level Management*, Ottawa, International Development Research Centre, 2002.

Bucks,D.A. & Nakayama, F.S. *Trickle Irrigation for Crop Production*, Elsevier, Amsterdam, the Netherlands, 1986.

Burman, R.D., Cuenca, R.H. & Weiss, A.: *Advances in Irrigation*, Orlando, Academic Press, Florida, USA, 1983.

Campbell, G.S.: *An Introduction to Environmental Biophysics,* Springer-Verlag, New York, 1977.

Charles, M. Burt: *The Surface Irrigation Manual*, Waterman Industries, California, 1995.

Chauhan, H. S.*: Elements of Micro-irrigation and Its Crop Application*, International Book, Delhi, 2007.

Chossat, J.C.: *Evaluation du Procede D'irrigation "Irrigasc",* Ministere de la Coopération et du Développement, France, 1992.

Copeland, M.C.: *A Manual for Irrigation Planning in Developing Areas,* Pretoria, South Africa, SECOSAF, 1993.

Dasberg, S. & Bresler, E.: *Drip Irrigation Manual*, Bet Dagan, Israel, Int. Irrig. Info. Ctr., 1985.

Devi, Sudharmai: *Analytical Procedures in Soil Science and Agricultural Chemistry*, Agrotech, Delhi, 2004.

Dinabandhu Sahoo: *Farming the Ocean: Seaweeds Cultivation and Utilization*, Aravali, Delhi, 2000.

Ghosh, G. K.: *Biopesticide and Integrated Pest Management*, APH, Delhi, 2009.

Gilley, J.R.: *Advances in Irrigation*, Orlando, Florida, Academic Press, USA, 1983.

Godden G.: *Growing Citrus Trees*, Australia, Lothian Publishing Company Pvt. Ltd., 1988.

Hanks, R.J.: *Efficient Water Use in Crop Production*, Madison, Agron, 1980.

Hardy B.: *Biology and Agronomy of Forage Arachis*, Cali, International Centre for Tropical Agriculture, 1994.

Hessayon, D. G. Dr.: *The Vegetable Expert*, England, PBI, Publications, 1985.

Hillel, D.: *Optimizing the Soil Physical Environment Toward Greater Crop Yields,* Academic Press, New York, 1972.

Hooja, Rakesh: *Irrigation Drainage: International and National Perspectives*, Agrotech, 2000.

Howard, A.D.: *Incised River Channels*, John Wiley & Sons, New York, 1999.

James, L.G.: *Principles of Farm Irrigation System Design*, Wiley, New York, 1988.

Jeffers P.: *Evaluation of Four Onion Varieties in Montserrat*, Plymouth, CARDI, 1992.

Kenneth H. Solomon and Allen R. Dedrick: *Standards Development for Microirrigation*, CATI Publication 950601, 1995.

Kenneth H. Solomon and Greg Jorgensen: *Subsurface Drip Irrigation*, CATI Publication, US, 1993.

Khan, Samiullah: *Plant Breeding Advances and in vitro Culture*, CBS, Delhi, 1997.

Koli, P.A. and A.C. Bodhale: *Irrigation Developments in India*, Serials, 2006.

Kovach, S.P.: *Injection of Fertilizers into Drip Irrigation Systems for Vegetables,* Univ. Florida, Gainesville, USA, 1984.

Kunelius T.: *Annual Ryegrasses in Atlantic Canada,* Ottawa, Agriculture Canada, 1991.

M Lakshmi Narasaiah: *Irrigation Management and Globalisation*, Discovery, 2006.

Macself, A.J.: *Soils and Fertilizers*, Satish Serial Pub, Delhi, 2005.

Meena Francis: *Biotech's Dictionary of Irrigation*, Biotech, 2005.

Merriam, J.L. & Keller, J.: *Farm Irrigation System Evaluation: A Guide for Management,* Utah State University, Logan, USA, 1978.

Mondal, R.C.: *Appropri-ate Technology,* Intermediate Technology Publications, London, 1974.

Monteith, J.L. *Applications of Soil Physics,* Academic Press, Florida, USA, 1980.

Murata, M.: *Development of Small-scale Irrigation using Groundwater Resources,* Water Dev., Zimbabwe and Inst. Hydrology, Wallingford, UK, 1995.

Narain, Vishal: *Institutions, Technology and Water Control: Water Users Associations and Irrigation Management in Two Large-scale Systems in India*, Orient Blackswan, 2003.

Narasimha Rao, P.: *Irrigation Development: Issues and Challenges*, Discovery, 2007.

Neetha N: *Institutional Choice in Irrigation: A Case Study of Distribution in a Command Area in Kerala*, Concept Pub, 2010.

Orson, W. Israelsen: *Irrigation: Principles and Practices*, Axis Books, 2010.

Petr, Tomi: *Fisheries in Irrigation Systems of Arid Asia*, Daya, 2007.

Phocaides, A.: *Technical Handbook on Pressurized Irrigation Techniques*, FAO, Rome, 2000.

Pradhan, Prachanda and Upendra Gautam: *Farmer Managed Irrigation Systems in the Changed Context*, Farmer Managed Irrigation Systems Promotion Trust, 2002.

Punmia, B.C.: *Irrigation and Water Power Engineering*, Laxmi Publications, Delhi, 2009.

Qassim, Abdi: *Sprinkler Irrigation*, Situation Analyse, IPTRID, UK, 2003.

Raju, K.S., Biere, A.W., Kanemasu, E.T. & Lee, E.S.: *Advances in Irrigation*, Academic Press, Florida, USA, 1983.

Ramann, E.: *The Evolution and Classification of Soils*, Asiatic Pub, Calcutta, 2006.

Rawitz, E. & Hillel, D.: *Drip Irrig.,* San Diego, California, USA, 1974.

Rimal Gautam, Suman: *Incorporating Groundwater Irrigation: Technology Dynamics and Conjunctive Water Management in the Nepal Terai*, Orient Blackswan, 2006.

Robinson, D.H.; *Entomology: Principles and Practices*, Agrobios, Delhi, 2001.

Rosegrant, M.W. & Perez, N.D.: *Water Resources Development in Africa*, Washington, DC, 1995.

Shalhevet, J.: *Irrigation of Field and Orchard Crops under Semi-arid Conditions*, Bet Dagan, Israel, Intl Irrig. Info. Ctr., 1976.

Sharma, M.L.: *Advances in Irrigation*, Orlando, Florida, USA, Academic Press, 1985.

Thomas F. Scherer at all: *Sprinkler Irrigation Systems*, MWPS, 1999.

Vaux, H.J., Jr & Pruitt, W.O.: *Advances in Irrigation*, Orlando, Florida, USA, Academic Press, 1983.

Withers, B. & Vipond, S.: *Irrigation: Design and Practices*, Ithaca, NY, USA, Cornell University Press, 1980.

Index

A

Agroeconomic Impacts, 216, 248.
Alluvial Fans, 110, 111, 116, 263.

C

Centre Pivot Irrigation, 5, 13.
Climatic Conditions, 7, 31, 41, 44, 82, 329, 330.
Comparative Advantage, 139.
Controlling Watercourses, 54.

D

Drainage, 26, 29, 30, 31, 33, 34, 35, 37, 38, 39, 42, 48, 49, 51, 74, 77, 93, 95, 96, 97, 98, 103, 104, 106, 113, 117, 118, 120, 121, 122, 123, 124, 125, 126, 127, 128, 129, 130, 131, 132, 133, 134, 138, 142, 152, 154, 156, 157, 158, 167, 169, 176, 210, 218, 221, 223, 234, 242, 244, 251, 253, 255, 262, 265, 267, 274, 275, 304, 309, 310, 311, 312, 320, 322, 328.
Drainage System, 34, 96, 120, 124, 127, 128, 130, 131, 132, 154, 156, 157, 158, 223.
Drainage System Design, 131.
Drip Farming, 9, 10.
Drip Irrigation, 5, 6, 7, 8, 9, 10, 11, 12, 13, 14, 15, 16, 17, 18, 19, 20, 39, 40, 41, 42, 43, 44, 48, 96, 105, 108, 109, 231, 265.
Drip Irrigation Systems, 6, 14, 15, 17, 20, 41, 109, 265.
Drip Irrigation Timer, 11.

E

Ecological Effects, 125.
Environmental Impact, 32, 96, 108, 304, 309.

F

Furrow Irrigation, 9, 37, 38, 47, 105.

H

Hydroelectricity, 53.

I

Influencing Irrigation, 102.
Irrigation Agency, 84, 86, 135, 140, 141, 144, 145, 146, 147, 148, 150, 151, 154, 155, 156, 157, 158, 159, 160, 161, 163, 164, 167, 170, 173, 175, 179, 181, 185, 186, 187, 188, 190, 195, 199, 201.
Irrigation Infrastructure, 66, 152, 168, 170, 175, 179, 197, 222, 223, 225, 226, 227, 228, 243, 319.
Irrigation Management, 63, 64, 133, 134, 135, 136, 143, 149, 153, 165, 173, 184, 185, 207, 209, 212, 239, 242, 253, 256, 259, 270, 271, 272, 273, 274, 281, 282,

287,   301,   330.
Irrigation Scheduling, 4,   5,   6,   7,
74,   75,   76,   78,   79,   82,   83,
84,   88,   89,   257,   281,   282,
283.

M

Modelling Reservoir Management,
55.
Modern Irrigation Techniques, 22.

N

National Policies, 177,   197.

O

Okavango, 111,   115,   116.
Organizing Processes, 159.

P

Paddy Irrigation, 216,   248.
Partial Rootzone Drying, 108.
Participatory Irrigation Management,   136,   144,   145,   172,
173,   199,   227.
Piping, 14,   49,   50.
Population, 59,   97,   98,   111,   113,
116,   117,   124,   126,   200,
201,   202,   204,   205,   208,
210,   211,   213,   214,   215,
291,   292,   306,   312,   313,
317,   318,   328,   329,   330.
Population Policy, 201,   202.
Practising Irrigation Scheduling, 84.

R

Reservoir, 33,   51,   52,   53,   54,   55,
67,   76,   85,   89,   98,   114,
246,   271,   314,   316,   317,
318.

River Water, 96,   98,   109,   110,
277,   279,   281,   310,   317.

S

Soil Permeability, 260.
Spate Irrigation, 49,   59,   60,   61,
62.
Surface Irrigation, 8,   36,   38,   39,
40,   80,   96,   105,   112,   223,
230,   286,   300,   304,   314.

T

Technical Requirements, 20.
Techniques, 1,   2,   5,   21,   22,   36,
44,   63,   91,   98,   99,   104,
110,   128,   160,   236,   267,
271,   278,   308,   309,   316.
Tidal Irrigation, 109.
Traditional Irrigation, 32,   92,   114,
207,   240.
Training Strategy, 150.

V

Vegetable Gardens, 19,   20.

W

Water Delivery Schedules, 63,   64,
65,   66,   67,   71,   72,   197.
Water Management, 17,   90,   200,
216,   234,   244,   248,   269.
Water management, 55,   57,   59,   63,
67,   84,   90,   91,   92,   101,
105,   115,   176,   187,   203,
205,   220,   224,   234,   236,
237,   238,   240,   241,   242,
243,   244,   252,   269.
Water Policies, 177,   240.
Water Source, 15,   19,   41,   45,   49,
50,   57,   278.

❑❑❑